Have courage to use your own understanding

Immanuel Kant, 1724-1804

Second Edition 08/2020

ISBN 978-3-9524178-4-3

Contents

1   Preface

Tunnels represent the most expensive type of transportation infrastructure, being more elaborate to construct than open roads, bridges and other civil engineering structures. The benefits must be worth the price. In comparison to an open road, tunnels usually provide a faster and safer means of transportation, shortcutting dangerous mountain or urban roads, protecting the road users from external hazards like bad weather, rockfall and avalanches, and in return shielding the environment from the negative impact of road traffic, like noise and pollution.

The tunnel ventilation system is an important element for the operational safety of a tunnel, contributing significantly to investment and operational costs. A lot of research on road tunnel ventilation has been worked out in the 20th century. Most of the findings are still valid. In the meantime, many experts have been retired or passed away, and taken their knowledge with them.

However, in the 20th century, the focus of tunnel ventilation was on air quality in the tunnel under traffic. Vehicle pollutant emissions were a serious problem. Since the late 1990[ies], air quality in and around road tunnels is rarely an issue, with some exceptions, since vehicle emissions have significantly decreased in developed countries with strict emission standards. Today, most road tunnel ventilation systems are never in operation. The focus of tunnel ventilation design and operation has shifted towards smoke control in the rare event of a fire in the tunnel. Unfortunately, in some catastrophic fire incidents, the tunnel ventilation significantly contributed to the disasters by fanning the fire and increasing the spread of smoke.

In the present Road Tunnel Ventilation Compendium, comprehension of basic aspects of tunnel ventilation, in the context of general safety of road tunnels, should be advanced. Tunnel ventilation fundamentals and simple, clearly understandable und practically oriented specifications for concept, design, realization and operation of road tunnel ventilation systems shall be described. This Compendium does not comprise aerodynamic calculations, since they can be looked up in corresponding technical literature (see bibliography). Instead, open questions, common errors and ambiguities in tunnel safety and ventilation issues shall be cleared, and design hints are given, focusing on best practice when safe and reliable functionality matters.

This Compendium stems from the authors extensive practical experience in the analysis, design, implementation, commissioning, testing, and operation of tunnel safety and ventilation systems for over two decades, which in turn is based on the knowledge of many competent colleagues and predecessors, and research worked out by several generations of experts.

2 Fundamentals

2.1 Risk reduction

A road tunnel is in first order a line of connection for traffic. For this, a ventilation system is not required a priori.

A risk, by definition, is the product of the extent of damage and the probability of a particular incident. The ventilation may be one of the means to reduce the operational risks which arise from the traffic of vehicles. Such risks are in first order traffic accidents, exposure to noxious substances in the tunnel and in the environment, fires and release of hazardous materials. These risks impose a threat for life and health of tunnel users and may lead to material damages and subsequent costs.

The tunnel ventilation system must be understood as part of a general safety concept, together with different educational, operational and technical safety measures, which focus on:

1. Prevention

2. Mitigation

3. Evacuation

4. Rescue

in order of importance and effectiveness (main layers of defense as specified in the TSI-SRT for rail tunnels [35]).

For risk assessment, not only the immediate danger to life and health of people is important, but also possible consequences of damages to the tunnel structure and resulting unavailability of the traffic link.

The biggest risk of road traffic in tunnels as well as on open roads are collisions. The first and most effective measure to reduce risk in a road tunnel are traffic restrictions, but such measures reduce the capacity of the tunnel along with the connecting road network[1].

A possible contribution of the ventilation in reducing the collision risk in road tunnels is assuring a sufficient clear sight distance. Particles from vehicle emissions, dust and tire abrasion make the air hazy and reduce the visibility. By that, the probability of incidents is increased. Therefore, threshold values for visibility impairment are defined which should not be exceeded during operation. In countries with restrictive vehicle emission standards, that is of importance only in case of long tunnels with bidirectional traffic. Especially in the portal zone, where the collision risk of the vehicles entering the tunnel is higher, the visibility should not be impaired.

---

[1] For example, after the fire incident on 17.10.2001 in the Gotthard Road Tunnel, unidirectional traffic for trucks was implemented as a first step and later the traffic was confined to a drip-feed system which resulted in the drop of the number of accidents by approx. 75%.

A special case is the fogging of windscreens of cars driving into a tunnel with bidirectional traffic during humid weather. Here, the collision risk can be substantially reduced through implementation of adequate ventilation, as explained in chapter 4.6.

Other aspects of the benefit of ventilation result from mitigation, i.e. the possible reduction of extent of damage. A health risk due to pollutant load in tunnels is very rarely an issue since the exposure time for passengers in a vehicle while crossing the tunnel is short. In practice, if the threshold values for visibility impairment are complied with, the pollutant load in most of the tunnels in countries with emission limits is negligible[2]. However, in case of exposure lasting over a longer time, for example during maintenance works, lower threshold values need to be adhered to.

The health of tunnel users, technical staff and emergency services in the tunnel environment should be protected by complying with the legal air quality limits. Furthermore, the ventilation can be applied to limit the increase of the pollutant load around tunnel portals due to contaminated air from the tunnel.

Fires in tunnels occur rarely, but in a statistical relevant frequency. However, the fire risk is usually much lower than that of collisions. The impact of ventilation on the fire risk results from the ability to influence the smoke propagation. Regarding the safety of tunnel users, the most important task is to improve the means of escape for the tunnel users. Emergency services can be supported by providing an easy access to the incident site by controlling the smoke propagation.

Fig. 1   In a tunnel filled with smoke

Risks arising from spillage of hazardous materials or explosions are practically not of great relevance for the tunnel ventilation due to low probability and lack of appropriate automatic detection possibilities but must be taken into account for the operational concept.

___

[2]In special conditions, for example at higher sea levels, pollutant emissions can substantially increase due to alternations of the combustion process in the engines.

The benefit of tunnel ventilation is a possible increase in safety by reducing the risks of collisions, exposure to pollutants and fires. Those benefits can only be judged keeping in mind other important influencing factors, especially the traffic characteristics, availability of shelters and safe escape routes, other safety equipment as well as the operational and safety concept.

Electrically driven cars impose a new challenge for fire safety, because a mechanical damage of a battery can result in a rapid fire development, which cannot be controlled by traditional water based firefighting.

Most important for traffic and tunnel safety are education, awareness and carefulness of drivers, but that goes beyond the scope of this Compendium.

2.2   Basic Tunnel Ventilation Principles

A tunnel is characterized by the fact that its dimensions in the longitudinal axis are larger in order of magnitude in comparison to other axes. As an approximation, tunnels are to be considered as a one-dimensional system under the aspect of longitudinal airflow in the tunnel tube, which is the most important parameter to consider.

Local three-dimensional effects have to be considered for complex geometric conditions, as in buildings and stations, but usually not in tunnels - with the exception of a possible smoke stratification in case of fire. Stratification occurs at well known, defined conditions as described in chapter 5.2.

Another important principle is that a tunnel is normally open at its ends to ensure free flow of traffic[3] and there is practically always a longitudinal airflow in the tunnel. In open tunnels, still-standing air does rarely, and only temporarily, occur.

A road tunnel is in first order ventilated by the moving traffic. Only in cases of congestion, tunnel closure or an incident, especially a fire, the traffic is supposed to stop and slow down the airflow. Beside the traffic, other factors determining the airflow are:

- Tunnel geometry
- Inertia of air
- Buoyancy forces (downwards or upwards) due to differences between inside and outside temperature
- Wind pressure on the portals
- Barometric pressure difference between the portals
- Changes in temperature and density (especially in case of fire)
- In case of fire: volume increase through the combustion process and heating and buoyancy due to fire heat release
- Operation of the ventilation system
- Operation of other systems in the tunnel (especially fixed fire fighting systems)

---

[3] There are tunnels which can be closed with gates, e.g. in case of mines, construction sites, and some rail tunnels. However, gates are applicable only in case of very limited traffic, and may impose an additional obstruction and risk.

These boundary conditions are essential for the design and operation of a ventilation system. As an example, in Fig. 2 the longitudinal flow velocities measured by different anemometers in a 2.5 km long road tunnel are displayed. The tunnel was closed between 20:15 h and 05:00 h; during that period, the flow was influenced only by external boundary conditions. The high fluctuations with changes in flow direction before 20:15 h and after 05:00 h were caused by the bidirectional traffic.

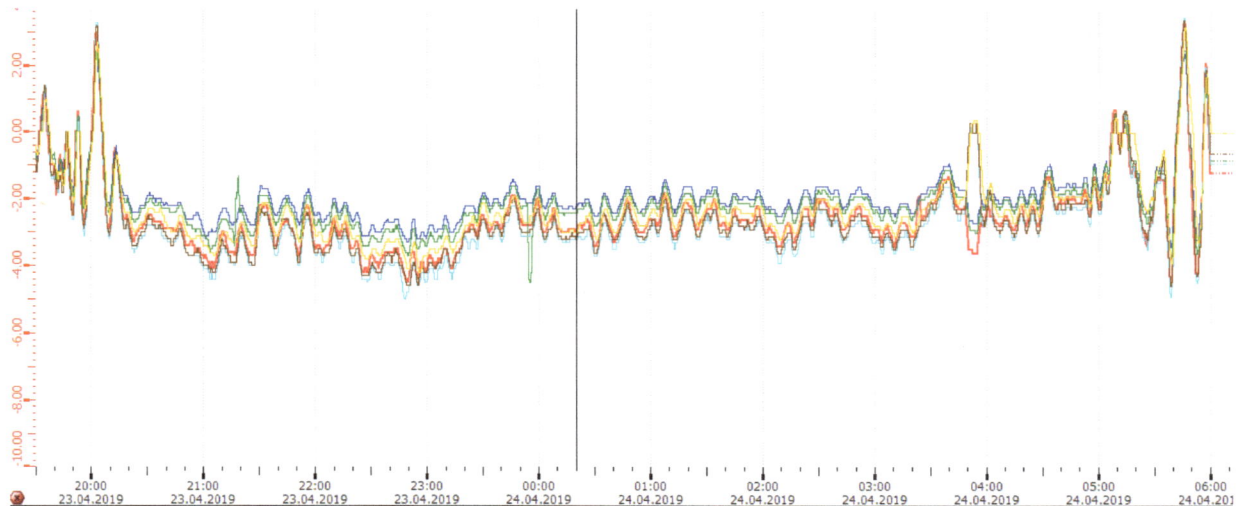

Fig. 2   Measured longitudinal flow velocities during tunnel closure; x-Axis: time, y-Axis: air speed [m/s] (copyright canton Uri highway administration)

2.3   Ventilation Goals

Goals for normal operation:

-     Sufficient sight distance in the tunnel and its portal areas for safe traffic operation.
-     Adherence to permissible threshold values for noxious substances for short-term exposure in the tunnel[4].
-     Adherence to emission limits in the tunnel environment (around portals and shafts).
-     Adherence to permissible threshold values for noxious substances for long-term exposure in the tunnel during maintenance.

Goals of fire ventilation:

-     Providing tenable conditions by controlling smoke propagation in the incident tube:
      1. To support escape of tunnel users
      2. To facilitate access to the emergency services[5]
-     Ensuring smoke-free shelters and escape routes

---

[4] Adherence to admissible climate conditions, e.g. temperature and humidity inside the tunnel, may be an issue in underground mass transport systems, not in road tunnels,

[5] The support of fire fighting operations should only be prioritized when the collapse of the tunnel structure due to a fire is possible and would lead to unacceptable consequences, e.g. in urban cut & cover tunnels.

2.4 Traffic and tunnel categories

For the conception, planning and operation of road tunnel ventilation, the traffic is the most important determining factor, being the source of risks in the form of vehicle emissions and fire loads, and carrier of people who are primarily affected by the risks. Moreover, the traffic significantly influences the airflow in the tube and hence, directly the ventilation.

For the determination and assessment of road tunnel ventilation concepts, the following categories are applied:

Category 0: Short tunnels

without ventilation system

In case of short tunnels, a mechanical ventilation system is not useful. According to most guidelines, a ventilation system is required only for road tunnels longer than 500 – 1000 m, depending on the traffic.

Category A: Tunnel with unidirectional traffic with free exit

Generally, multiple tube rural highway tunnels

In tunnels category A, it is important to avoid blockage of vehicles by appropriate traffic measures, taking into account also the connecting road network. In case of exceptional incidents, which could lead to a traffic jam, the affected tube needs to be closed and the traffic in the connecting road network should be accordingly managed to assure free traffic flow in the tunnel.

Fig. 3   In a category A highway tunnel

Category B: Tunnel with unidirectional traffic with possibly blocked exit

Generally, multiple tube urban tunnels with high traffic load

Critical for the categorization of a tunnel with unidirectional traffic is exclusively the question whether, in case of an incident, the vehicles can leave the tunnel downstream of the incident site or whether they get trapped. For that, the probability of congestions and traffic management in the connecting road network need to be considered. In category B tunnels chances of traffic jams are supposed to be high. For the activation/control of a suitable fire ventilation program, a reliable detection should assess whether at the time of fire alarm, vehicles in the tunnel are trapped or not.

In both Category A and B tunnels, the airflow is in traffic direction, as long as there is sufficient traffic. Only in situations with little traffic, for instance at night, or during a congestion after an incident, external forces may lead to a flow reversal.

Category C: Tunnel with bidirectional traffic

Generally, single tube tunnels

In category C tunnels, the airflow direction may vary (see Fig. 2). In case of a fire, it is supposed that blocked vehicles and people may be situated on both sides of the incident location.

Fig. 4   Example of category C tunnel: Tunnel du Mt. Blanc in 1965

Special categories

Transitions between closed tunnels and open roads need to be treated cautiously, for instance short open road segments between consecutive tunnels, or open sided galleries.

3 Ventilation concepts

The ventilation concepts are in principle defined by the direction of airflow in relation to the tunnel longitudinal axis.

3.1 Continuous longitudinal ventilation

In case of longitudinal ventilation, the direction of the airflow in the tunnel is parallel to the tunnel longitudinal axis. A continuous longitudinal ventilation is the simplest system in which the air flows in through one portal and out of another portal. A continuous longitudinal airflow persists in all tunnels when the ventilation system is switched off.

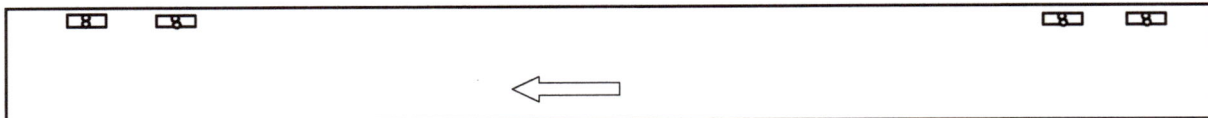

Fig. 5   Continuous longitudinal ventilation

3.2 Sectional longitudinal ventilation with exhaust and/or supply

A tunnel with longitudinal ventilation can be divided into several sections by means of concentrated extraction (of polluted air / smoke), or concentrated injection of supply air in the tunnel.

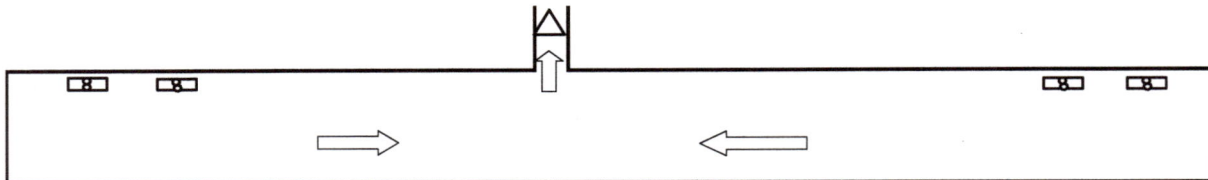

Fig. 6   Fixed concentrated extraction, with longitudinal airflow in both branches towards the extraction

In many tunnels with bidirectional traffic, fixed concentrated extraction is applied in the middle area of a tunnel. For emission control in urban tunnels with unidirectional traffic, a concentrated extraction may be required near the exit portals.

In case of very long tunnels, several extraction and injection points at different locations are possible. These are especially common in metro and rail tunnels in the form of 'push-pull systems', but usually not applied in road tunnels.

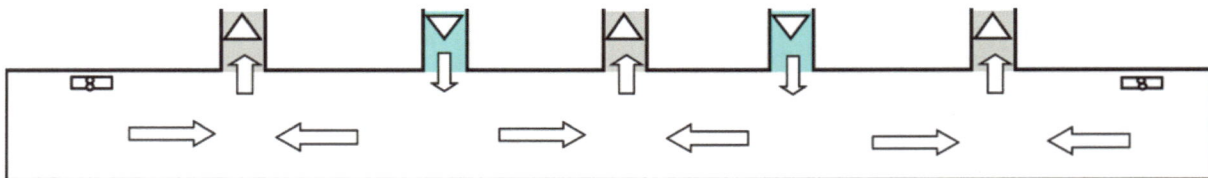

Fig. 7   Fixed concentrated extraction and injection, with longitudinal airflow in sections (push-pull)

Note that in push-pull systems, the flow direction changes in every section.

In very long tunnels with unidirectional traffic, an air exchange system with extraction and adjoining fresh air injection with impulse in traffic direction can be useful to limit longitudinal air velocities during high fresh air demand.

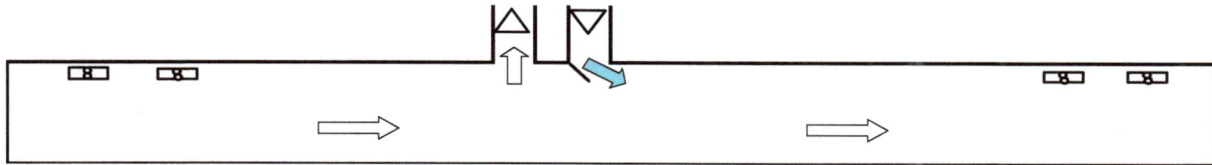

Fig. 8   Longitudinal ventilation with air exchange stations

With an extraction through remote controlled dampers and a parallel exhaust duct, the extraction can be adjusted at a desired location. This concept is suitable for long tunnels with bidirectional traffic and high traffic load and can be used for normal operation as well as in case of fire. Despite of high investment costs for ducts and fan stations, this is the applied standard ventilation concept in many modern Alpine tunnels with bidirectional traffic.

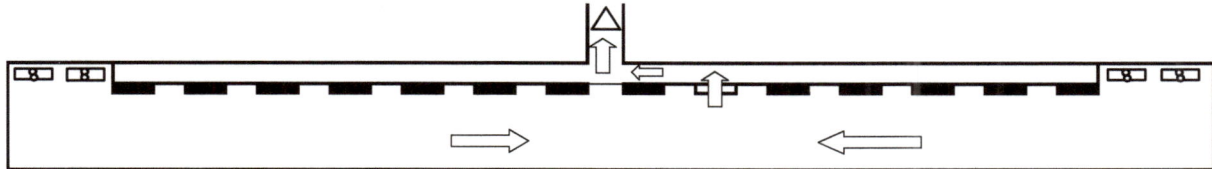

Fig. 9   Controllable concentrated extraction, with longitudinal airflow in both branches

3.3   Single vs. multiple point exhaust

To understand the effect of an exhaust, the flow towards the extraction point is crucial. In a tunnel without fresh air injection, that's the linear flow from the portals in both sections, as in Fig. 6.

Simultaneous multiple point extractions slow down the airflow between the extraction points, and in case of symmetrical extraction, the air might even stand still in that area. There is no exchange of air; smoke and contaminant concentrations would peak, see Fig. 10. This happens also in semi-transversal ventilation systems, see Fig. 12, Fig. 13.

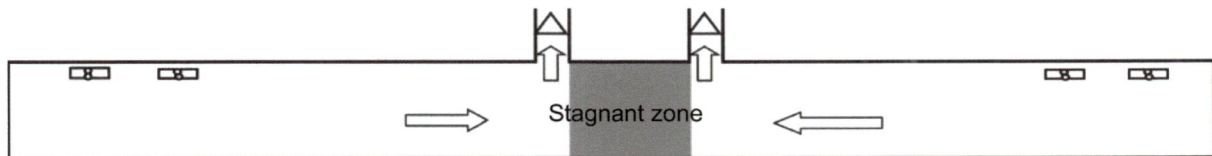

Fig. 10   Flow conditions with simultaneous extraction at two points

To extract the air between two points, fresh air would have to be supplied between the extraction points as displayed in Fig. 7. The problem is that by blowing air into the exhaust zone, where the fire is supposed to be, the smoke would be locally swirled and the fire might be additionally fanned. That's why even in transversal ventilation systems, where fresh air supply would be available (see chapter 3.4), the supply must be switched off in the fire zone upon a fire alarm.

Distributed smoke extraction may be usefully applied for instance in buildings, or subway and underground rail stations, where the required supply flow is enabled from the open station or through open emergency exit doors.

In practice, damper size in road tunnels is often limited, to not exceed a certain flow velocity in the damper and to limit the resulting pressure difference between the tunnel and the exhaust duct. Therefore, multiple dampers have to be opened to achieve the required exhaust volume, despite the mentioned disadvantages. In that case, it is essential to ensure a longitudinal flow between the open dampers, by adjusting the parameters of the flow control accordingly.

3.4   Transversal ventilation

In case of transversal ventilation, the main flow direction is perpendicular to the longitudinal tunnel axis. The transversal flow is achieved by a linear supply air injection and by linear exhaust, both distributed along the length of the tunnel.

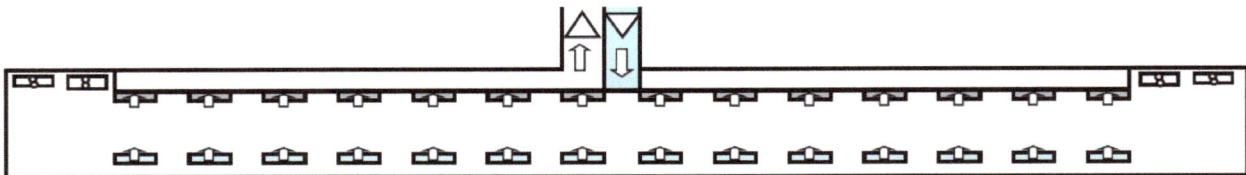

Fig. 11   Transversal ventilation (with jet fans in portal zones)

Transversal ventilation was originally developed for the 2,6 km long Holland tunnel under the Hudson river in New York, which was opened in 1927, and became the worldwide standard ventilation system for many other tunnels built in the following decades, before vehicle emission restrictions became effective in the 1990[ies]. In very long tunnels with bidirectional heavy traffic, and in countries with high vehicle emissions, transversal ventilation systems are still useful.

Originally, fixed openings along the tunnel for air injection and exhaust were foreseen. Today, remote controlled dampers are used for exhaust. During normal operation, all dampers are inclined at a small opening angle. In case of fire, the transversal ventilation is switched to controllable concentrated extraction at the fire location with longitudinal ventilation in both branches (Fig. 9), by opening the dampers close to the fire and closing all other dampers.

Transversal ventilation slows down the longitudinal airflow velocity in the tunnel tube. In tunnels with bidirectional traffic, this is an advantage for the controllability as well as for the smoke propagation in case of fire, particularly before the fire detection. On the other hand, the disadvantage is the high energy consumption. For the control of the longitudinal flow, particularly for fire ventilation with concentrated extraction, additional jet fans are required.

In the case of unidirectional traffic, transversal ventilation is not useful because it hampers the natural ventilation induced by the traffic, which would otherwise be sufficient to achieve the air quality and visibility requirements.

3.5   Semi-transversal ventilation

Semi-transversal ventilation occurs when omitting either supply air or exhaust air in a transversal ventilation system. In comparison to transversal ventilation, significant investment and operating costs can be saved using semi-transversal ventilation. Depending upon the setting of supply and exhaust air quantities, any combinations between transversal and semi-transversal ventilation are possible[6].

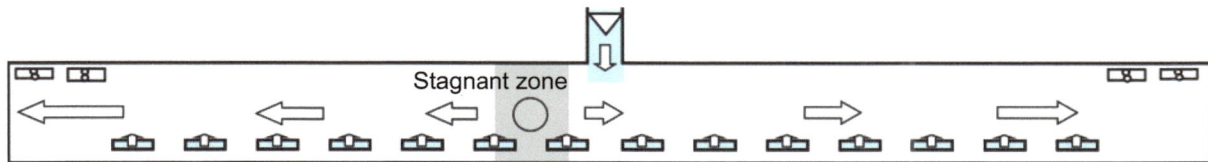

Fig. 12   Supply semi-transversal ventilation

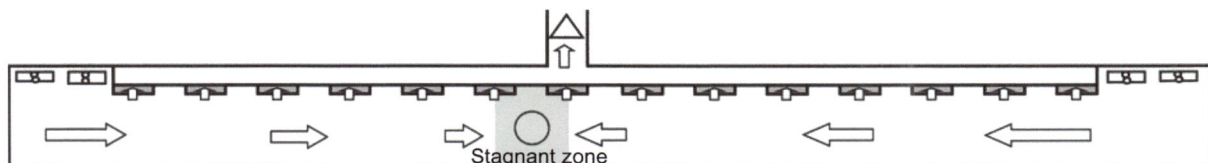

Fig. 13   Exhaust semi-transversal ventilation

In case of semi-transversal ventilation, the transversal flow transforms into longitudinal flow and the longitudinal flow velocity increases towards the portal areas. Supply semi-transversal ventilation has a linear injection evenly distributed over the length and the airflow in the tunnel is from inside to outside.

On the other hand, exhaust semi-transversal ventilation has a linear extraction evenly distributed over the length with airflow in the tunnel from outside to inside. During operation of semi-transversal ventilation, a point of stagnant air builds up in the tunnel (see Fig. 12, Fig. 13). At this point no exchange of air takes place and pollutant concentration can increase there even while the ventilation is running. This phenomenon can be observed in form of local zones with poor visibility. In practice, this point of stagnant air and high concentration is blurred due to diffusion and turbulence caused by moving traffic. Moreover, this point is not stationary but moves longitudinally in the tube.

Semi-transversal ventilation always induces a longitudinal airflow in the tunnel. This flows outwards progressing from the point of stagnant air towards the tunnel portals in case of supply semi-transversal ventilation and in case of exhaust semi-transversal ventilation slowing down from the tunnel portals towards the point of stagnant air.

---

[6]For the concept of controllable concentrated extraction as in Fig. 9, sometimes the term 'semi-transversal ventilation' is used by civil engineers, referring to the exhaust duct. From the ventilation point of view, this is technically incorrect. During operation of concentrated extraction, longitudinal ventilation prevails in two sections with flow towards the extraction point.

In the road tunnel fires in the Mont-Blanc, Tauern and Gotthard road tunnels as described in chapter 8.17, smoke propagation was intensified due to the longitudinal flow induced by the operating semi-transversal ventilation. From today's point of view, semi-transversal ventilation systems are not suitable for tunnel ventilation, since they can lead to high local pollutant concentrations and on the other hand induce locally high flow velocities. In the past, this disadvantage was often accepted to reduce the high investment and operating costs of transversal ventilation systems.

3.6   Longitudinal flow control

Longitudinal flow control is essential for all kinds of ventilation systems – with or without exhaust - for fire ventilation, and in special cases may be required for emission control during normal operational.

Without flow control, a determined ventilation operational state results in a mostly random flow situation, which depends not only on the operation of the ventilation system, but on random boundary conditions, that is the forces as specified in chapter 2.2.

For further explanations on longitudinal flow control, see Lit. [38], [91]. Basis of the flow control is a precise, reliable flow measurement in the tunnel. The momentum for the longitudinal flow control is usually applied by jet fans without injection or impulse nozzles with concentrated, directed point injection in an acute angle to the tunnel axis. Jet fans have a better efficiency and are nowadays common for tunnel ventilation applications. However, in special cases, impulse nozzles can be advantageous.

Independent of the installed ventilation system, an intentional basic control of longitudinal flow is possible by means of mobile jet fans.

Important is the fact that the jet from impulse fans or nozzles requires a certain distance for transfer of momentum to the tunnel air, which is in the range of dozens of meters. When switching on an impulse fan, in the first seconds a backflow occurs, which generates a pressure, slowly accelerating or decelerating the tunnel air movement. The operation of impulse fans in a smoke layer would immediately distribute the smoke over the whole cross section.

Using push-pull, transversal or semi-transversal ventilation systems with multiple ventilation sections, longitudinal flow control is also possible, but only to a limited extent. Other installations, for example air curtains or mechanical curtains, are generally not useful for tunnel ventilation applications.

The longitudinal flow can be suppressed by closing the tunnel with gates or curtains, but this is not implementable in most road tunnels, since they need to be open for moving traffic.

There are different levels of flow control:

1. No longitudinal flow control

In tunnels without any ventilation system or where the ventilation system is switched off, there is no active longitudinal flow control.

2. Fixed setting

With a fixed setting, a predetermined number of fans is operated in a defined direction, without taking the real flow conditions into consideration. In reality, a wide range of flow conditions can occur depending upon the momentary boundary conditions.

Fixed settings for flow control are suitable for the normal operation ventilation, when admissible concentrations of noxious substances must not be exceeded, or levels of visibility maintained, but the longitudinal air velocity does not matter.

For fire ventilation, a fixed setting can be counterproductive and even worsen the situation, since the real airflow is a random value and can reach high speeds, depending on the boundary conditions as explained in chapter 2.2. From today's point of view, fixed settings without flow control according to measured flow velocities are not acceptable for fire ventilation, with the following exceptions:

- As fallback modes in the case of failure of the flow control / flow measurement
- In non-incident tubes of twin tube tunnels
- In short tunnels of category A, where closed-loop flow control is not feasible due to spatial restrictions for the placement of anemometers and jet fans.
- In the second phase, when no endangered people are left in the tunnel, to support the emergency services by providing a smoke-free access to the incident site and blow the smoke out of the tunnel.

3. Staggered closed loop control

In staggered closed loop flow control, jet fans are switched on and off step-by-step based on the measured airflow velocity.

The staggered control process is quite slow and therefore can balance only asymmetric conditions and slow fluctuations in long tunnels with high inertia of the moving air, but not rapid airflow fluctuations, Overshooting, in extreme cases even a flow reversal, must be avoided. Therefore, the flow control cannot be activated before the traffic slows down and stops, which can last many minutes in long tunnels.

After activation of flow control, it takes take a certain time until the desired state is achieved, depending on initial conditions. This time delay, however, may be in the same range as the expected time for self-evacuation, see Fig. 21.

The advantage of a staggered control, in relation to a continuous control, is reduced complexity, since no VSD are required. From today's point of view, staggered closed loop flow control may be acceptable for:

- low risk category A tunnels
- tunnels with concentrated smoke extraction and large exhaust quantities, when the flow control serves only to support the exhaust
- emission control (portal exhaust)

4. Continuous closed loop control

In a state-of-the-art continuous closed loop control, PID or more highly developed controllers, for instance Model Predictive Control (MPC), are implemented. The impulse fans are equipped with variable speed drives (VSD).

To control the longitudinal airflow, knowledge of the actual flow condition in the tunnel is essential. A closed loop control of longitudinal airflow requires a complex instrumentation and control process which may potentially increase higher error rate as well as the amount of effort required for the implementation and tests. Stringent criteria apply for the precision and reliability of airflow velocity measurements, which provide the input for the controller (see chapter 8.10), along with regular functionality tests and calibration of the measuring instruments.

The closed loop flow control shall fulfil the following requirements:

1. **The target state needs to be achieved quickly.**
   The target value of airflow velocity should be achieved within a defined bandwidth in shortest possible time after activation of the closed loop control.
2. **An overshooting, or even flow reversal, must not occur.**
   After the target value was achieved for the first time, the airflow velocity must remain within the bandwidth. Flow reversal must be prevented.
3. **Deviations are to be quickly corrected.**
   In case of activated closed loop control, the airflow must achieve the target value within the bandwidth in shortest possible time after the occurrence of a defined deviation.

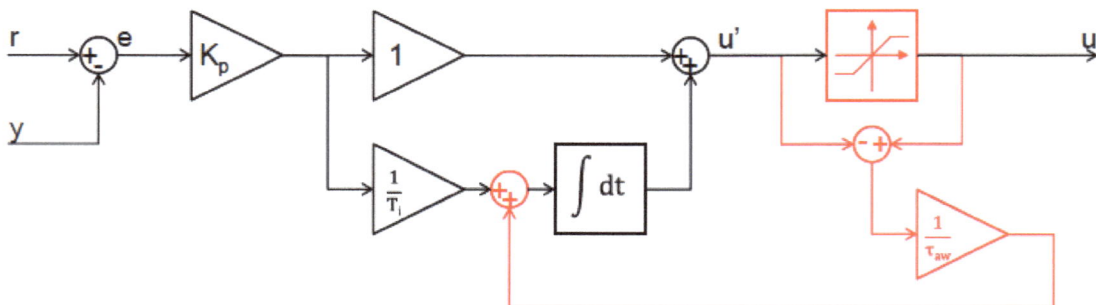

Fig. 14   Controller schematic (from [38])

4 Ventilation for normal operation

4.1 Visibility and air quality in the tunnel

The air quality in the tunnel is primarily affected by the vehicle emissions and dust. It is assumed that the exposure time of the passengers in the tunnel is very short. Therefore, the threshold values for pollutants to be adhered to, especially for particles, CO, NO_x which are applied for tunnel ventilation design, are not valid for long-term exposure. Particles hinder the sight, increasing the collision risk and lowering the possible driving speed, hence limiting the traffic capacity. In practice, in most road tunnels, the opacity is the decisive factor for the normal operation ventilation.

When the first long road tunnels with mechanical ventilation were built in the 1920[ies], high levels of vehicle emissions were to be expected. The main task of the tunnel ventilation was to restrict the air pollutant concentrations and visibility impairment in the tube. For the design of normal operation ventilation, the stationary fresh air demand was decisive. The air quality requirements in most of the longer tunnels were met with transversal ventilation systems and respective high operating costs. This was the case until the early 1990[ies].

For stationary traffic conditions, the fresh air demand to limit the air pollutant concentrations and opacity values to the permissible threshold values can be easily calculated. The relevant vehicle emission factors and calculation models are defined in many planning guidelines and in the PIARC publication [86].

Nowadays, in developed countries with strict vehicle emission control and increasing number of electrically driven cars, the stationary fresh air demand is not relevant for the assessment and design of ventilation for most tunnels and therefore its calculation is unnecessary. Dust from tire abrasion may be more influential than vehicle emissions. Either no normal operation ventilation is required at all, or the fire ventilation layout covers the requirements of normal operation ventilation. Only in very long tunnels with high traffic load, the design for normal operation may become an issue. Rather, dynamic aspects of ventilation need to be considered.

4.2 Displacement und dilution

Air pollutant concentrations and opacity can be reduced by means of two different mechanisms:

- Displacement of polluted air through fresh air in case of longitudinal ventilation
- Dilution of polluted air by fresh air in case of transversal air supply

In practice, the two mechanisms mix up more or less. As a matter of principle, air quality in wide areas of the tunnels is better in case of displacement and therefore in case of longitudinal ventilation, as compared to that in case of dilution.

4.3 Dynamics of airflow

With the ventilation system switched off, there is a continuous longitudinal flow in the tunnel[7]. In the case of constant flow direction and distributed sources of air pollutants (i.e. vehicles) in the tunnel, air pollutant concentrations increase in the direction of the airflow and reach their maximum at the exit portal.

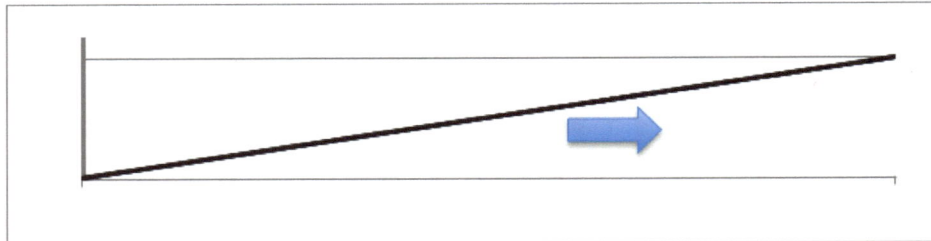

Fig. 15   Concentration profile in case of constant flow direction

In tunnels with unidirectional traffic, this concentration profile is generally constant. Only the amplitude of the pollutant concentration peak varies depending upon the emissions in the tunnel and the flow velocity. The lower the flow velocity, the steeper is the concentration ramp.

In tunnels with bidirectional traffic, the airflow direction can change, thereby shifting air pollutant concentration peaks from the tunnel portal to the inside of the tunnel. These concentration peaks are visible in form of local zones with poor visibility in the tunnel.

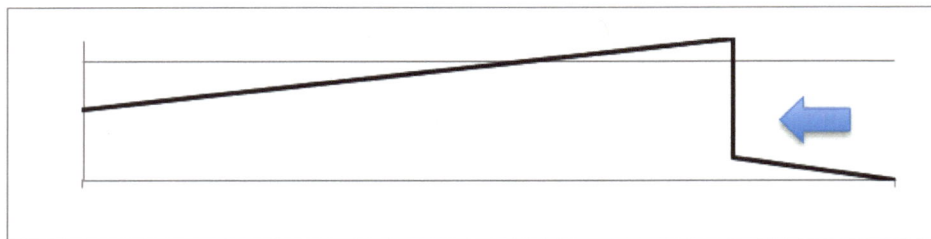

Fig. 16   Schematic concentration profile after reversal of flow direction

A flow reversal can occasionally also take place in a tunnel with unidirectional traffic and under strong meteorological pressure against low traffic. This is however not relevant in practice because of low traffic and hence low emissions in that situation.

In reality, the concentration peak is blurred by diffusion and turbulence created by the moving vehicles and travels longitudinally in the tube. In tunnels with bidirectional traffic the concentration peak never stays still, since most driving forces of the longitudinal airflow as listed in chapter 2.2, especially the traffic, are dynamic. In case of balanced traffic in both directions, the concentration peak moves back and forth, as the traffic is never completely symmetrical.

---

[7] Except in case of additional injection or extraction, when using a shaft with natural buoyancy.

In case of long lasting asymmetrical traffic distribution or strong external pressures, a constant flow direction and thus the concentration profile as shown in Fig. 15 occurs. In short tunnels, the concentration peaks are for that reason pushed quickly towards the portals.

Opposed to short tunnels, in longer tunnels under unfavorable conditions, the concentration peaks are unable to leave the tunnel. For example, in case of high traffic volume with high emissions and rather balanced traffic in both directions the concentration peaks move back and forth and tend to increase. To avoid exceeding the opacity or pollutant threshold values, the forced ventilation must then be switched on.

After switching on the longitudinal ventilation, concentration peaks are shifted towards the portals approximately with the speed of the airflow. This takes some time and during this process, the longitudinal ventilation must keep on operating in the given direction. The dynamic behavior of the traffic is generally faster than that of the air in the tunnel and moreover, it is unforeseeable. If during the operation of the ventilation the main traffic direction changes before the concentration peak reaches the portal, the longitudinal ventilation must act against the traffic. Therefore, a mechanical longitudinal ventilation in bidirectional tunnels should be able to achieve a minimal airflow velocity against the main traffic direction for an assumed traffic distribution in the two directions.

It needs to be considered that in a tunnel with bidirectional traffic when the air pollutant concentration is at its maximum at the tunnel portals as in Fig. 15, the vehicles coming from outside enter the tunnel in an area with more or less turbid air which increases the collision risk.

When applying a concentrated extraction, the concentration peak can be extracted from the tunnel near its actual position, provided that a reliable detection is available.

4.4   Limits of longitudinal ventilation

A high fresh air demand leads to a high airflow velocity in a tunnel with longitudinal ventilation. But the flow velocity needs to be limited:

- The power demand of the installed ventilation system and hence its energy consumption while operating increases with the third power to the flow velocity.
- High flow velocities raise dust and particles, impairing visibility in the tunnel.
- High flow velocities lead to rapid smoke propagation in the tunnel in case of fire. This is disadvantageous especially before the fire is detected and the fire ventilation would start to operate.

Therefore, a threshold value for the flow velocity in the tunnel is defined. If that limit would be exceeded by operation of the ventilation system, continuous longitudinal ventilation is not feasible.

Possible measures to reduce the flow velocity are:

- Traffic restrictions (as a temporary measure).
- Increase the tunnel cross section8
- Air exchange stations (as per Fig. 8) in long category A and B tunnels with unidirectional traffic.
- In category B and C tunnels, sectionalized longitudinal ventilation with concentrated extraction (as per Fig. 6 or Fig. 9). In long tunnels additional openings (e.g. entrance and exit ramps) or fresh air injection inside the tunnel for a sectionalized longitudinal ventilation in multiple sections.
- Transversal ventilation (as per Fig. 11) in long category C tunnels.

4.5   Normal operation ventilation control

The goals of the normal operation ventilation control are:

- The permissible opacity values and air pollutant concentrations in the tunnel and its environment should not be exceeded
- Optimizing the energy consumption of the ventilation system
- Stable fan operation

In countries with limitations on vehicle emissions, the normal operation ventilation is usually needed only in long tunnels with bidirectional traffic (Category C) or in congested city tunnels (Category B). In the latter, the demand for emission control may require the operation of the ventilation system to prevent air from the tunnel exiting the portals.

In most other tunnels, the ventilation system operates only in exceptional cases, e.g. when the traffic comes to a standstill due to an accident. Then, the energy consumption is negligible. Normal operation ventilation might be controlled by inexpensive low resolution opacity meters / smoke detectors, whereby the tunnel needs to be ventilated within a short time with high ventilation power. By applying hysteresis and delay times, frequent switches and load changes must be avoided.

Based on the annotations in chapter 4.3, it is clear that in case of longitudinal ventilation, operating of only few jet fans or providing low exhaust quantity is of little use. When there is a demand for operation of the ventilation system, a lot of traffic in the tunnel is to be expected, which is emitting pollutants and particles, and also significantly influences the airflow. A long tunnel with bidirectional traffic and longitudinal ventilation can be effectively ventilated only with high power consumption.

In tunnels where the ventilation system is expected to run often, fine-tuned controlling makes sense to save energy costs. The lead value of the controller may then be defined usually by high resolution opacity meters, by the traffic surveillance system or alternatively by time programs.

---

8 Being a very costly measure for long tunnels, this is usually not feasible

In tunnels with extraction through remote controlled dampers, the extraction point is determined by the position of the opacity meter (or measuring device for noxious substances) with maximum measured value. After the extraction has been started, the extraction point remains fixed till the measured value falls below a defined threshold value.

In case of regular operation, it is worthwhile to optimize the energy consumption. In this aspect, transversal ventilation can be controlled much easier and more precisely as compared to longitudinal ventilation, which reacts slowly due to the tunnel air inertia.

A closed loop control is normally not required for the normal operation. One of the exceptions may be emission control in a tunnel in which no polluted air may exit the portals due to emission restrictions. Especially in tunnel networks, complex controller systems with stringent requirements might be required.

4.6   Fogging of windscreens

When warm, damp air is cooled down quickly, moisture will condensate on cold surfaces. In tunnels with bidirectional traffic, in cold and humid weather, vehicles entering the tunnel from outside may get fogged windscreens. This, at a sudden, impairs the visibility of the driver and can lead to serious accidents.

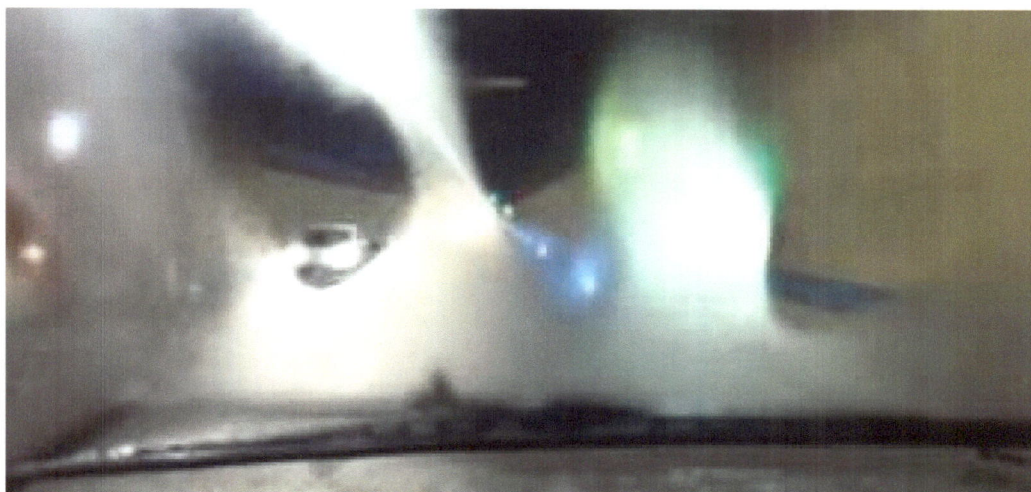

Fig. 17   Fogged windscreens (public domain)

In road tunnels, the combustion engines of the vehicles produce heat and moisture. Depending on the weather, additional moisture is brought in by rain and snow. In long mountain tunnels, rock heat and water lead to a further increase in temperature and humidity. In case of continuous longitudinal airflow, the air flows in through one portal, is heated up and humidified, and flows out from another portal. At the exhaust portal, the warm, humid tunnel air will condense out on cold surfaces or particles in the cold ambient air.

Correspondingly, the cold windscreens of vehicles driving against the flow direction of effluent air get fogged up. This occurs only in tunnels with bidirectional traffic since in tunnels with unidirectional traffic, no vehicle should enter the tunnel from the exit portal. Vehicles entering the tunnel in the same direction as the airflow are surrounded by ambient air, and air temperature and humidity increase only gradually while driving along the tunnel.

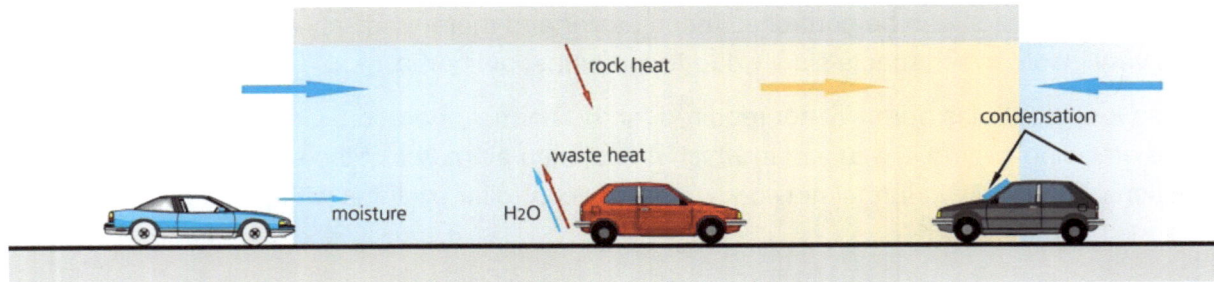

Fig. 18   Humidity problems at outlet portal

In practice, the oversaturation of the air can be used as reference value for the assessment of humidity problem. Oversaturation is the difference between the absolute humidity inside and the maximum absolute humidity on the outside, to be calculated from the measurements of outdoor air temperature, indoor air temperature and relative indoor air humidity. If threshold values of oversaturation are exceeded, the windscreens of the vehicles entering the tunnel may fog up.

Fogging of windscreens can be avoided when the airflow direction is from outside into the tunnel at both portals. To achieve that, the exhaust must be in operation. Under consideration of asymmetries through traffic and meteorological forces, high exhaust quantities are required and/or the flow would need to be controlled additionally by jet fans. This increases energy consumption.

Fig. 19   Flow and temperature pattern during extraction in operation

4.7   Emission control

With emission control ventilation, it may be avoided that the contaminated tunnel air exits from the tunnel portal, especially in urban environments. Rather, the air is extracted from the tube and is vertically blown out with a high velocity through a shaft and diluted in the atmosphere. Alternatively, the exhaust air may be filtered.

Emission control ventilation can improve the pollutant load in the proximate surrounding of the tunnel but has no measurable influence on the global pollutant load, not even when using filters. Filter systems though have psychological advantages over exhaust shafts.

Fig. 20   Exhaust shaft for emission control at the portal of an urban highway tunnel (public domain)

Exhaust systems, and possibly filters, may be designed for the complete extraction of an expected airflow in the tunnel, plus a safety margin to ensure an airflow from outside into the tunnel at the exit portal. On the other side, the flow control in the tunnel must be operated to not exceed the exhaust airflow. By that, emission control ventilation in operation leads to a high energy consumption. From the point of view of total ecological balance, using emission control ventilation adds to the burden on the environment[9].

---

[9] Since the underlying emission predictions in Environmental Impact Assessments (EIA) are mostly too conservative, many emission control ventilation systems are never in operation.

5 Fire Ventilation

5.1 Goals and principles

Fires in tunnels may impose a serious threat. While the heat close to the fire location may impair and eventually destroy the infrastructure and equipment, the smoke affects the people, even far away from the fire. People in the smoke may suffer poisoning, lose orientation, and eventually lose consciousness and die due to toxic combustion products. As a simple principle, there are no tenable conditions in the smoke.

The decisive questions for defining fire ventilation concepts are:

- Is protection of people or protection of the infrastructure given a higher importance?
- Which flow conditions are to be expected under different circumstances?
- Where would endangered people be located and what means of escape do they have?
- Where and how will emergency services, especially the fire fighters, get access?

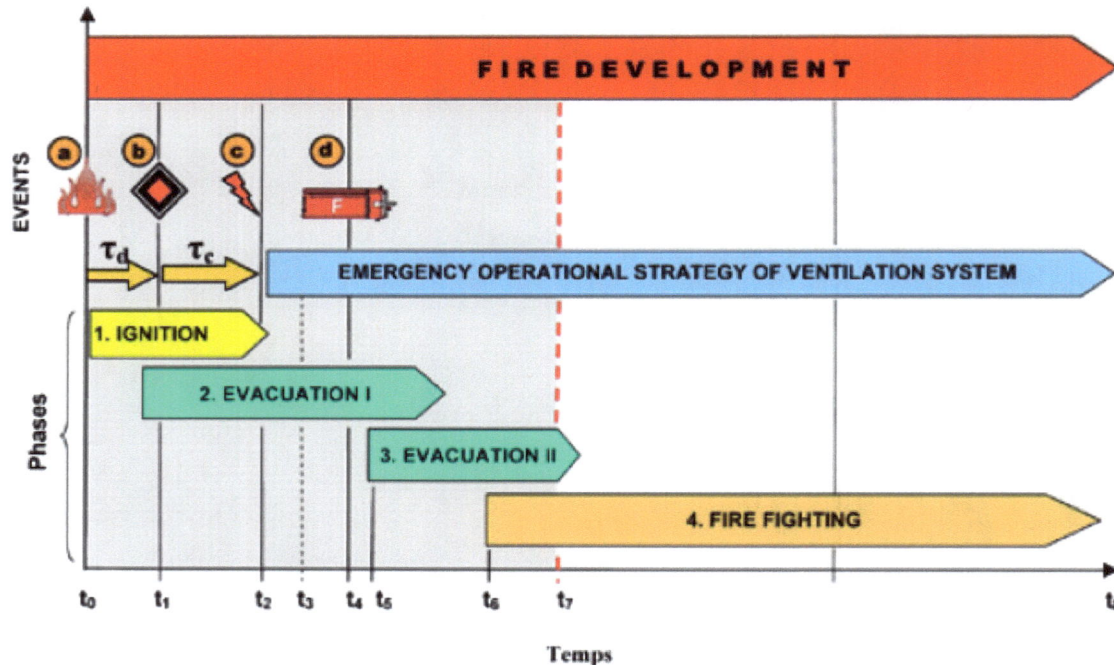

Fig. 21   Phases of a fire incident (from [79])

Unlike in closed buildings, longitudinal flow always occurs in a tunnel, and may lead to a quick smoke propagation in case of fire, putting the health and life of people situated in the path of smoke in danger. Still-standing air as an initial condition is a rare exception. The smoke propagates along the tunnel in one or in the other direction or in both directions simultaneously before the incident is detected and the fire ventilation and other safety measures, especially stopping the traffic, come into effect. The initial condition for fire ventilation is in most cases a tunnel already filled with smoke moving over a long distance.

Fire ventilation in tunnels means in first place controlling the longitudinal airflow.

Without forced ventilation, the smoke moves predominantly by meteorological forces, which cannot be controlled, by the traffic as long as it keeps moving, and the fire dynamics. In steep tunnels, the chimney effect by the fire can get very strong and drives the smoke quickly upwards. But fast smoke spread can also be caused deliberately by the operation of the ventilation system. This works fine as long as there are no people exposed to the smoke on the other side, where it is blown. Unfortunately, pushing the smoke towards one side has been often performed without considering where endangered people may be. That resulted in approximately 300 fatalities in the metro Baku (Azerbaijan) in 1995, and many fatalities in road tunnel fires as described in chapter 8.17.

In the following description, four different principles of fire ventilation are to be differentiated by increasing complexity:

-     Natural flow - no forced ventilation (Chapter 5.3)
-     Longitudinal smoke control with high flow velocity (Chapter 5.4)
-     Longitudinal smoke control with low flow velocity (Chapter 5.5)
-     Smoke exhaust - Concentrated extraction (Chapter 5.6)

5.2   Smoke stratification

An important goal of fire ventilation in buildings is the facilitation and retention of smoke stratification. People under a smoke layer can escape and are not directly at risk. Therefore, retaining a possible smoke stratification should be a fire ventilation goal in a tunnel too, particularly where people may be trapped at both sides of the incident site.

In practice, smoke stratification does not always occur in tunnels. For smoldering fires with low temperatures, smoke stratification is not to be expected. Moving traffic and high flow velocities prevent smoke stratification, especially before the fire is detected and the safety systems are activated, stopping the traffic and starting the ventilation with longitudinal flow control. Further, turbulences from impulse fans and activation of fixed fire fighting systems like sprinklers or fog systems destroy any smoke layering.

Smoke moving along the tunnel is cooled down by the tunnel walls and ceiling. This gradually decreases the temperature difference between the smoke layer and the underlying cold air, and finally the smoke layering ceases. By then, the smoke is distributed over the whole cross section. At still standing air, the smoke would come down at both sides of a fire. To prevent that, a minimal flow velocity must be maintained to allow for a one-sided smoke removal, as explained in chapter 5.5.

Fig. 22   Smoke movement in a tunnel with still-standing air.

Still standing air is achieved by closing the tunnel. In closed tunnels, an airflow would have to be generated by additional supply or exhaust, as is common for mine or underground site ventilation systems.

Smoke layering is always locally limited in the range of approximately several dozen meters around the fire location. In this zone, people must have the possibility to escape. Therefore, in tunnels with longitudinal ventilation without exhaust, shelter or emergency exits should be provided in short distances.

Smoke stratification in tunnels is supported by the following measures:

-   Stopping the traffic, since moving vehicles, in particular large trucks and busses, may destroy the smoke layer.
-   Not operating impulse fans in the smoke layer.
-   Slowing down the longitudinal flow to a low value.
    Benefit: This helps in slowing down the smoke propagation, ideally to a lower value than the assumed escape speed of people.
-   Avoiding flow reversal.
    Benefit: In case of fire in a tunnel, it is to be assumed that people do not immediately escape and stay where there is no smoke. In case of a flow reversal, they would be exposed to the smoke.

Fig. 23   Smoke stratification at controlled flow conditions

In the following cases, retention of smoke stratification is not the primary goal:

- When operating fixed fire fighting systems (FFFS)
  (generally prioritizing the protection of the infrastructure).
- In case of concentrated smoke extraction.
- Fire close to a tunnel portal when the flow is to be reversed.
- In the second phase to provide one-sided access to emergency services.
- In other applications than road tunnels, e.g. when the ventilation goal is smoke dilution.

Most importantly, longitudinal smoke control with high flow velocity as described in chapter 5.4, i.e. exceeding the so-called critical velocity as required by many standards, does impede smoke stratification and hence leads to non-tenable conditions downstream of the fire.

5.3 Natural flow – no forced ventilation

Goal: Not to disturb the natural smoke propagation (and possibly smoke stratification)

Natural flow is useful for short tunnels without ventilation system, or in tunnels without reliably working closed loop control. It is better to leave the ventilation system turned off instead of possibly worsening the situation, when no information on the emergency situation and the state of flow in the tunnel is available[10].

This case can also occur accidentally in all tunnels in case of control failure or power supply failure. For a scenario analysis, such malfunctions need to be considered accordingly.

5.4 Longitudinal smoke control with high flow velocity

Goal: Move smoke towards one direction without backlayering

Historically, this was the standard fire ventilation principle in tunnels category A with unidirectional traffic, where smoke propagation in traffic direction was strived for. It was based on the assumption that the vehicles should be able to safely leave the tunnel between the incident site and the exit portal, and no people would be situated downstream of the fire. The people who are stuck in front of the incident site are situated in a smoke-free, safe area.

Fig. 24   Longitudinal smoke control without backlayering and no people downstream

---

[10] In fact, before a reliably working flow control was developed, the most useful fire ventilation scenario in longitudinally ventilated category C tunnels without local extraction was to switch off the fans.

To avoid backlayering, the so-called critical velocity must be exceeded. Unfortunately, a design to achieve the critical velocity against adverse conditions leads to higher flow velocities in operation for less adverse conditions, especially when using only a fixed setting without control. In direction of smoke movement, the smoke is distributed all over the cross section, and conditions become very dangerous. Smoke spreads quicker than people can escape. High flow velocities also lead to additional fanning of the fire and increased heat release rate.

In reality, it cannot be assured that no people are situated downstream, even in category A tunnels. Actually, high flow velocities have caused most of fatalities in tunnel fires. On the other side, there is no reason to not allow backlayering. Therefore, the paradigm of critical velocity for fire ventilation must be given up, with the exception of flow reversal in portal zones as described in chapter 5.8. As a consequence, the closed-loop control of the longitudinal airflow with a low target velocity as described in chapter 5.5, allowing for backlayering, is proposed as a standard solution for fire ventilation in most tunnels.

Forcing the smoke towards one direction without backlayering may be a requirement for firefighting operations as described in chapter 5.10. Smoke control with high flow velocities may be useful in other applications than road tunnels:

-    In rail/metro tunnels with a narrow cross section, where smoke stratification is unlikely.
-    When the fire size can be limited, e.g. by use of exclusively non-combustible materials.
-    When the smoke can be diluted to an extent that allows for tenable conditions, and high flow velocities would not fan the fire

5.5   Longitudinal smoke control with low flow velocity

Goals:

-    Move smoke towards one direction and avoid flow reversal
-    Slow down smoke propagation below assumed escape velocity of people
-    Retain possible smoke stratification

Fig. 25   Longitudinal smoke control with low flow velocity

With longitudinal ventilation, the tunnel is always smoked-up in one direction or even in both directions. To support the escape of tunnel users, the longitudinal flow must be stabilized at a low flow velocity. A flow direction reversal would not be acceptable, even when the pressure conditions change.

Smoke stratification may be enabled, and backlayering of the smoke layer has to be accepted. However, if the stratification is stable, this poses no risk to human life.

A low flow velocity is also desirable to support the efficiency of fixed fire fighting systems, even when smoke stratification cannot be achieved in that case.

Limiting the flow to a target velocity, within a certain bandwidth, requires a closed loop control. By that, the target flow condition can be achieved in a few minutes, if designed and applied appropriately. Today, this is the recommended standard approach for most longitudinally ventilated tunnels.

5.6   Concentrated smoke extraction

Concentrated extraction enables an airflow on both sides towards the point of extraction, drawing back the smoke. Smoke free zones on both sides of the extraction point allow a safe stay of tunnel users and emergency services. This increases the safety especially in long tunnels with heavy bidirectional traffic. Thus, the goals of a concentrated extraction for fire ventilation are:

-    Limiting smoke propagation in the tunnel
-    Creating smoke-free zones on both sides of the fire zone

Concentrated extraction is useful particularly in long category C tunnels and in special cases in category B tunnels with blocked exit, were people are supposed to be located at both sides of an incident site. That is achieved by controllable dampers in an exhaust duct along the tunnel, which are opened close to the fire location. In tunnels with transversal ventilation, the switching between distributed and concentrated extraction is done by adjustment of controllable dampers.

When available, concentrated extraction should be used not only for fire ventilation, but also to ensure an adequate visibility and air quality during normal operation by locally extracting pollutants from the tunnel.

In some older tunnels, fixed extraction points were installed. The exhaust should be operated in case of fire near the extraction point. Through fixed extraction points, the tunnel is conceptionally divided into short tunnels with longitudinal ventilation. For fire locations in the sections between the extraction points, the longitudinal ventilation should be controlled as described in chapter 5.5.

The dynamic behavior of the smoke movement before and during the exhaust has to be understood:

-     At the time of fire detection and starting of the fire ventilation, the smoke has already spread with the longitudinal flow in the tunnel. Depending on the detection time and initial flow velocity, the smoked-up area may stretch over many 100 meters. There is no smoke stratification, the smoke has cooled down and been distributed over the whole tunnel cross section (see Fig. 22).
-     After start-up of the extraction, the direction of flow is reversed at one side of the tunnel, i.e. where the smoke has mainly spread before the start-up of extraction. The cold, distributed smoke is pulled back, and a possible smoke stratification there is destroyed. By that, the conditions, which may have been tenable before the start of the extraction, will deteriorate.
-     Pulling the smoke back towards the point of extraction is a dynamic process until a stationary smoke containment is achieved. This may take several minutes even under ideal conditions (see Fig. 26). By then, most of the endangered people have either escaped from the vicinity of the incident site or are not capable of escaping.

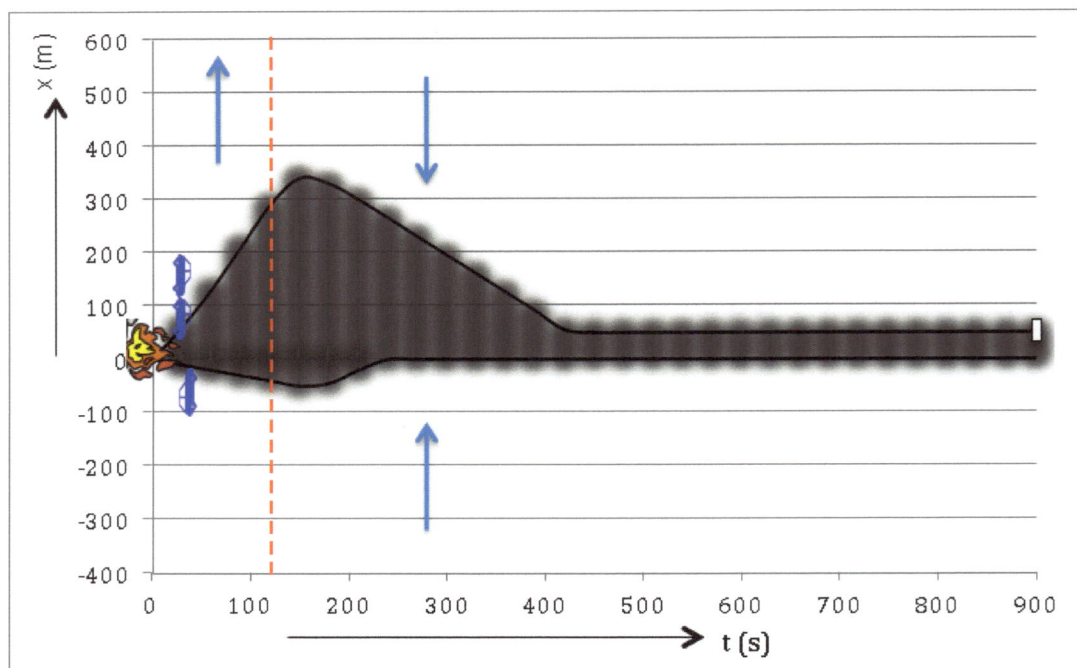

Fig. 26   Dynamic smoke propagation with an ideal extraction through one open damper, fast detection (2 Min.), fast flow control

- Therefore, the smoke extraction has a very limited benefit, and may have even a detrimental effect, in the first phase for the self-rescue of people in the immediate vicinity of the fire location.
- Only after the smoke has been pulled back and a steady state is achieved, the smoke extraction can contain the smoke in the zone of stratification
- Thus, smoke containment is of great help in safeguarding people further away from the fire and supporting the emergency service teams in the second phase.

Fig. 27   Concentrated extraction close to the fire – the target steady state of the scheme in Fig. 26

Further arguments to be considered:

- The ideal state without smoke propagation in the tunnel, with a concentrated smoke extraction directly above the incident site would be subject to coincidence, when the fire location is close to an exhaust damper. Usually, at least the section between the fire location and the open damper is filled with smoke.

Fig. 28   Front of smoke area under outermost open exhaust damper

- When dampers further away from the fire location are opened, the smoke is spread over a longer section of the tunnel. In this case, the extraction may worsen the situation. This happens in practice in case of wrong detection, or erroneous operation.
- A closed loop control of longitudinal flow as described in chapter 3.6 is required additionally to the exhaust system. Without control of longitudinal flow, the smoke could not be extracted from the tunnel in many scenarios.
- The lower the exhaust volume, the more stringent are the requirements to the closed loop control and the longer it takes until the smoke is contained. With a strong exhaust system, the requirements to the control of longitudinal flow become less important.
- On an increase of the exhaust volume, the power requirement increases with third power (at same boundary conditions), or the exhaust ducts must be accordingly largely dimensioned, which again increases the constructional investment costs.
- In comparison to longitudinal ventilation systems, concentrated extraction leads to significant additional costs, especially for the construction of exhaust ducts and fan buildings, the large exhaust fans and the necessary power supply.
- By providing an exhaust duct over a slab in the tunnel ceiling, the space for a possible smoke stratification in the tunnel ceiling is reduced.
- A slab in the ceiling leads to an additional risk since it may collapse.

Fig. 29 Collapsed slab in the Sasago Tunnel in Japan, 2012 (public domain)

- As described in chapter 3.3, a concentrated extraction should ideally be realized through one open damper as near as possible to the fire location. By opening additional dampers, the situation gets worse, since the area between the open dampers remains smoked-up for a longer period of time. In fact, that is where most of the endangered people are expected to be.
- The layout with a high exhaust volume flow through one open damper leads to large dampers and elaborate constructional provisions. In practice, many exhaust systems are designed for multiple open dampers. By opening additional dampers, the exhaust volume flow can be slightly increased and as a result, the flow in the tube would be accelerated faster towards the exhaust zone. The flow control must be adjusted accordingly, to enable a longitudinal flow between open dampers.

5.7 Smoke extraction in tunnels with unidirectional traffic

The following arguments have to be considered:

- In tunnels with unidirectional traffic, concentrated extraction is useful when vehicles and people get trapped downstream of the incident site. However, even without extraction, those people have a means to reach an emergency exit and escape through the cross passages to the other tube.

Fig. 30   Longitudinal smoke control and vehicle trapped downstream of the fire

- In case of continuous longitudinal ventilation, the fire ventilation does not change the flow direction and can be started immediately. With extraction, the flow direction downstream of the fire is changed.
- The dampers must be opened downstream of the incident site. When opening the dampers upstream of the incident site, e.g. due to false detection or due to human error, the area where blocked cars and persons are located would be smoked-up.

Fig. 31   Unidirectional traffic and concentrated extraction upstream of the fire, endangering people

- In category A highway tunnels, the possibility of incorrect use of extraction (Fig. 31) is more probable than trapped passengers downstream of the incident site (Fig. 30). As a result, concentrated extraction leads not only to substantial additional costs, but also may increase the risk for the tunnel users.
- In urban category B tunnels, this risk may be judged based on scenario analyses under consideration of possibilities of traffic management.
- An advantage of concentrated extraction in tunnels with unidirectional traffic may be the possible mitigation of constructional damages and therefore a reduction of costs and time for refurbishment after major fire incidents. However, for that purpose, fixed fire fighting systems might be more useful.

5.8 Flow reversal in portal zones

Portal zones represent a different situation, since outside of the tunnel, the smoke from the tunnel will be diluted and tenable conditions are to be expected. Thus, the exterior area is perceived as a safe zone. It is to be assumed that most people near the portal have secured themselves by immediately escaping out of the tunnel. In this case, the primary task of fire ventilation is to avoid smoke propagation towards inside the tunnel.

Airflow outwards in the direction of the portal will mostly occur at the exit portal of twin tube tunnels with unidirectional traffic, and in single tube tunnels with bidirectional traffic due to the influence of vehicles travelling inside the tunnel in the direction of the incident site. Then, the fire ventilation principle is the same as in the tunnel: Flow control without flow reversal according to chapter 5.5.

However, airflow and therefore smoke propagation from the portal into the tunnel is to be expected at the entry portal of twin tube tunnels with unidirectional traffic, and in single tube tunnels with bidirectional traffic when unfavorable external forces prevail over the traffic influence. Even smoke from a fire outside may be sucked into the tunnel. To prevent that, it is usually recommended to reverse the flow as quickly as possible and blow the smoke outward.

In the zone where flow reversal has occurred, the smoke will be distributed over the whole cross section and escape conditions deteriorate. No smoke layering is to be expected. Therefore, the definition of a portal zone where flow reversal is strived for is based on assumptions that:

- endangered people could escape to the outside
- the smoke has not spread too far yet at the moment of detection
- the fire location has been reliably detected close to the portal

Flow reversal in a portal zone defines one of the demanding design cases in all tunnels, also in tunnels with concentrated extraction, according to chapter 7.1.

Flow reversal is also the target condition in the non-incident tube of a twin tube tunnel, see Fig. 34.

5.9 Short tunnels, galleries and open roads

From the above-mentioned goals and fire ventilation concepts, the limits of usefulness of fire ventilation in short tunnels become evident:

- (As described in chapter 4, ventilation is not required for normal operation)
- In short tunnels, the number of people potentially exposed to threats is low; escape routes to the portals are short.
- In case of fire, short tunnels would be smoked-up across a large part before the ventilation could even react.
- Control of longitudinal flow in short tunnels is not feasible; there is no space for the necessary jet fans and anemometers at adequate distances.
- A smoke extraction in a short tunnel would be an option, but hardly justifiable from a cost-benefit point of view. In some cases, its operation could even deteriorate the conditions, as explained in chapter 5.6.
- In the second phase, to provide access to the fire department, longitudinal smoke control towards one side can be achieved with the help of mobile jet fans which are brought in place on request.

In short highway tunnels, the ventilation should focus on the non-incident tube, whereas in the incident tube, a flow control is not feasible.

Galleries, which are partly open, cannot be considered to be completely open from a ventilation point of view. Real fires and simulations have shown that the smoke may flow over long stretches in the open gallery and the visibility there is almost as low as it would be in closed tunnels.

Fig. 32   Smoke spread from an open gallery (copyright canton Graubünden road administration)

Short open stretches between consecutive tunnels on the same route also cannot be ignored, since the tunnel air, and in case of fire, the smoke, which flows out of one portal, can spread into the adjacent tunnel. Whether and to what extent this may occur, depends upon the surrounding terrain and on the wind conditions.

5.10 Supporting fire-fighting operations

After completion of the self-rescue phase, moving the smoke towards one direction without backlayering is often required by the fire department. That way, fire fighters can work in smoke-free environment and without respiratory protective equipment. Especially in short tunnels without ventilation equipment, they may apply mobile jet fans for this purpose.

From a safety point of view, in tunnels with a functioning closed-loop flow control, an override of automatic fire ventilation by fire fighters is usually not recommended. It can never be assured that no people are endangered in the direction of the smoke. Moreover, safe firefighting operations are feasible in stable conditions at low velocities below a smoke layer.

Fig. 33   Firefighting a HGV fire under a smoke layer at controlled flow conditions
(copyright fire department Schwechat)

On the other hand, after the fire, when being sure that no endangered people are situated in the tunnel, simply blowing the smoke out of the tunnel may be adequate.

5.11 Fire ventilation operation

Reliable, fast incident detection and a quick start of fire ventilation contribute significantly to the benefit of ventilation. The start-up dynamics of fire ventilation should be examined based on simulations right in the design stage and must be tested on site after completion.

In practice, it often takes a long time until the desired condition is achieved. This time may be in the same magnitude as the duration of the self-rescue phase (see Fig. 21). Particularly in case of low traffic and therefore less affected people in the tunnel, the benefit of ventilation is reduced mainly to the second phase when providing access to emergency services.

The initial state in tunnels is generally a switched off ventilation system with continuous longitudinal flow resulting from the traffic and the meteorological boundary conditions. Only in long tunnels with heavy traffic, an initial state with operating ventilation is likely. Then, the possibility of an incident increases and more people are potentially affected. The traffic that still moves after the start of the fire influences significantly the airflow.

In tunnels with unidirectional traffic, the direction of airflow for fire ventilation is the same as during switched off ventilation or during normal operation, namely in the direction of traffic movement. In case of longitudinal airflow against the direction of traffic movement, little traffic and thus a lower risk is to be assumed. This scenario may occur due to strong meteorological forces acting against the traffic direction.

Cases of unidirectional traffic with congestion (Category B) or bidirectional traffic (Category C) and continuous longitudinal ventilation are problematic. For the normal operation ventilation, high flow velocities may be required to ventilate the tunnel when high emissions occur. Conversely, in case of fire ventilation, the flow velocity must be slowed down. For this, it is important to be able to differentiate between smoke from a fire and opacity due to dust and vehicle emissions during normal operation, which is not easy in practice.

In tunnels with smoke extraction, an exhaust damper is to be assigned to a particular fire zone and smoke detector. That damper should be opened. It may be useful to open additional dampers in the smoked area in the first phase of extraction and to close them again if the respective smoke detectors do not show any signs of smoke anymore. All other dampers must be closed. When the dampers within a specified distance from the incident location cannot be opened due to some error, the dampers can remain closed and the control should proceed as in case of continuous longitudinal ventilation as described in chapter 5.5.

In tunnels with bidirectional traffic and concentrated extraction, the exhaust should also be used for the normal operation ventilation. By that, the fire ventilation concept is equivalent to the normal operation ventilation concept. After a fire alarm, the extraction point needs to be adjusted according to the detected fire location.

Most importantly, jet fans in the area around the fire location, where smoke stratification may occur, must be blocked. Therefore, the exact detection of the fire location is essential.

In twin-tube tunnels, the ventilation in the non-incident tube has to ensure that an overpressure is built up and smoke is prevented from spreading to the non-incident tube at the portals or by the cross passages. Generally, this is achieved when the jet fan groups next to the portals are operated with blowing direction towards the tunnel interior and other jet fans in the same direction as the airflow in the incident tube, i.e. against the traffic direction in the non-incident tube.

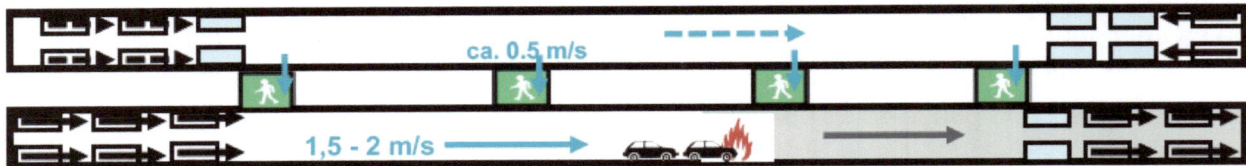

Fig. 34   Fire ventilation in both tubes of a highway tunnel (see also Fig. 35)

5.12 Traffic management

Accident and fire risks can significantly be reduced by limiting speed and / or the number of vehicles in the tunnel. By that, the benefit-cost ratio of any other safety measures, in particular the tunnel ventilation, decreases.

In category A highway tunnels, congestions in the tunnel happen only rarely. In case of exceptional incidents which can lead to a traffic jam, the affected tube needs to be closed and the traffic in the connecting road network should be accordingly managed until the free flow of traffic in the tunnel can be reassured. In category B urban tunnels, the chances of traffic jams are supposed to be high, because the traffic management in the adjacent road system cannot ensure that vehicles are able to leave the tunnel in case of an incident.

Since the traffic significantly influences the longitudinal flow in the tube, fire ventilation in road tunnels doesn't work without traffic control. Appropriate signaling must ensure that after the alarm, no more vehicles drive into the tunnel, but the vehicles which are located in the traffic direction behind the detected fire location can freely drive out of the tunnel.

In all twin-tube tunnels, also the non-incident tube must be immediately closed in case of a fire alarm, since otherwise, in case of moving traffic, the goals of fire ventilation, especially the prevention of smoke spread at the portals, cannot be fulfilled.

Enforcement of driver's compliance to signals is crucial.

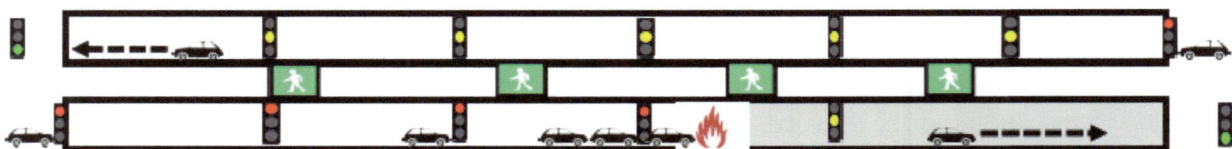

Fig. 35   Fire alarm traffic signaling in both tubes of a highway tunnel

6 Ventilation of escape routes

6.1 Fundamentals

In case of an incident, the traffic space and the tunnel portals are the most important and often the only means of escape from the tunnel. In twin-tube tunnels, cross-passages between the tunnels serve as escape routes. In single-tube tunnels, appropriate shelters or separate escape routes, for instance in form of parallel service tunnels with cross passages may be foreseen[11]. Such escape routes help the tunnel users to escape from the danger zone, and help the emergency services like police, ambulance and fire department, when direct access through the tube is hindered. Furthermore, they can also be used as access routes by the maintenance staff for inspections and service works during traffic.

Fig. 36   Emergency escape door in a road tunnel (example with sliding door)

Shelters and escape routes are generally separated from the tube through doors to provide a safe zone. Most importantly, the tunnel users must be able to open these escape doors under all circumstances. To guarantee the opening of doors, the pressure difference across the doors must be considered. These pressure differences arise from meteorological boundary conditions, traffic and the operation of the ventilation systems in the tunnel and in the shelter / escape route.

On using wing doors, this pressure difference has to be strictly limited, which is often not possible. There are many cases known in practice, where the escape doors could not be opened during the operation of fire ventilation in the main tunnel and running overpressure ventilation in the escape route. In such cases door opening assistance with servo drives are recommended. Sliding doors are in this regard less sensitive and therefore more suitable for the application as escape doors in tunnels.

---

[11] According to the European directive [7], shelters without access to the open are not allowed. From a cost-benefit point of view, this requirement must be questioned.

6.2   Prevention of smoke entry

The ventilation should keep the escape routes free of smoke as far as possible.

Entry of smoke from the tube into a separated shelter or escape route is prevented when the doors are closed or when airflow towards the tube prevails in the opening, in case the doors are open. To maintain the required airflow, an overpressure in the escape route is needed.

In case of an airlock, when one door is open and the second one behind it is closed, entry of smoke cannot be completely prevented, but can at least be restricted to a great extent, since no continuous airflow can take place in the open doors. A restricted entry of smoke is possible only through turbulence based on local effects.

In twin-tube tunnels, cross-passages should either be closed through one door or through two doors in series, building an airlock.

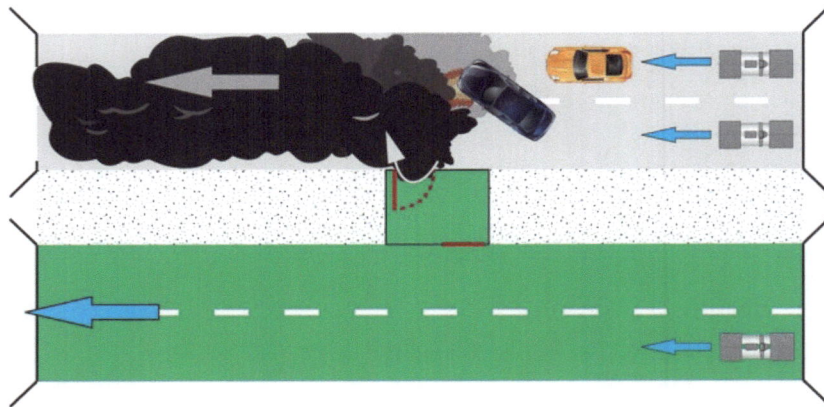

Fig. 37   Cross-passage with two doors providing an airlock

In case of fire, usually the jet fans in the non-incident tube are applied to achieve an overpressure against the incident tube. In case of open escape doors, by that it is ensured that smoke cannot spread to the non-incident tube, which represents a safe zone.

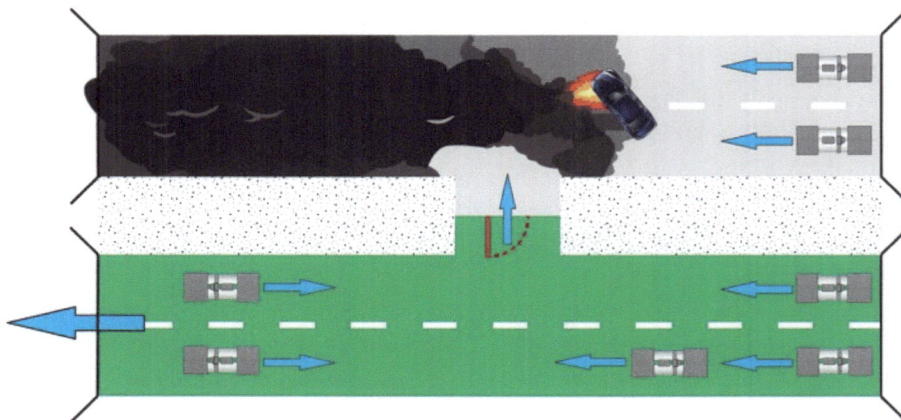

Fig. 38   Cross-passage with open escape door and ventilation in non-incident tube
                generating airflow through the open doors

Though the term 'overpressure' is common, important is the airflow through the openings. Relevant standards define a minimal airflow velocity to be achieved in open doors to prevent smoke entry. Local conditions need to be kept into consideration. Injection of an air jet in the opening can lead to a flow reversal, which in fact would suck the smoke into the escape route.

The pressure difference over the door should be limited so that the escape doors can be opened. At the same time, the operation of tunnel ventilation in both tubes is to be considered with meteorological pressures for different scenarios.

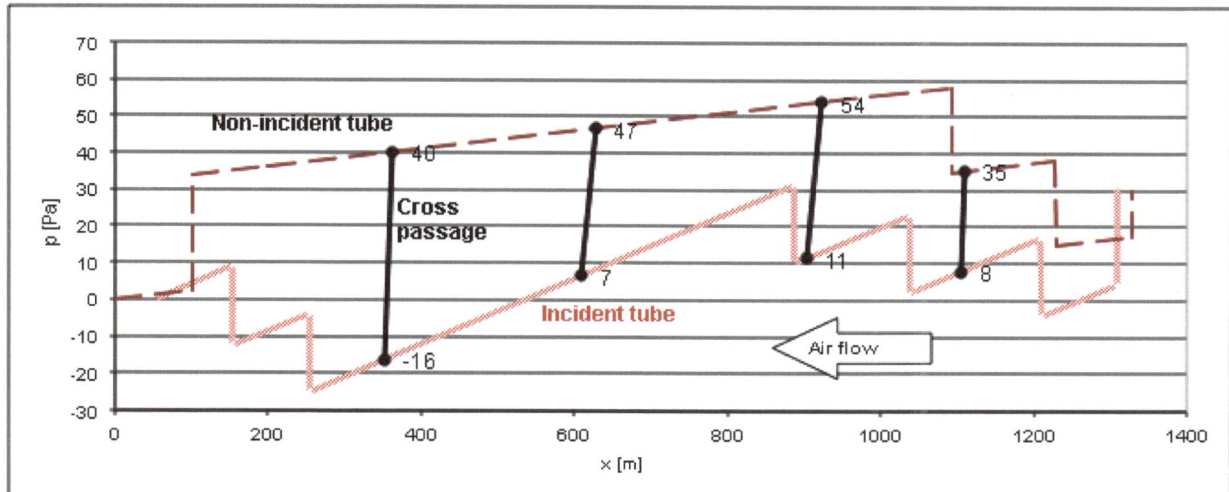

Fig. 39   Example: Simulation of pressures in both tubes during operation of fire ventilation

Possible measures in case of excessive pressure differences would be:

- Abandoning of overpressure and using an airlock
- Replacing wing doors with sliding doors
- Installing overpressure dampers towards the tube[12]
- Active pressure control
- Powered door opening

Overpressure ventilation from separate escape routes is initially started using an alarm or by opening an escape door. The airflow should be built up immediately after the door opening to prevent entry of smoke. At the same time, it is to be considered that the impulse required for the acceleration of air propagates at the speed of sound. In long tunnels with overpressure fans at the portals, it would take several seconds before the air starts to flow.

If sensible devices are placed in the escape route, permanent overpressure is to be foreseen to prevent polluted tunnel air from entering. For maintaining overpressure during normal operation, the volume flow can be reduced as compared to the nominal flow during an incident.

----

[12] Application of overpressure dampers or pressure control is only possible in case of overpressure ventilation of separate escape routes, but not in cross-passages in twin-tube tunnels.

In case of short escape routes without requirements to a permanent overpressure, purging at periodical intervals generally suffices.

For the ventilation of separate escape routes, it should be considered that the pressure difference over the escape doors to the tube is not equivalent to the pressure difference between escape route and outside, except in case of very short tunnels. Therefore, overpressure dampers on the outside not always lead to the required pressure relief over the escape doors.

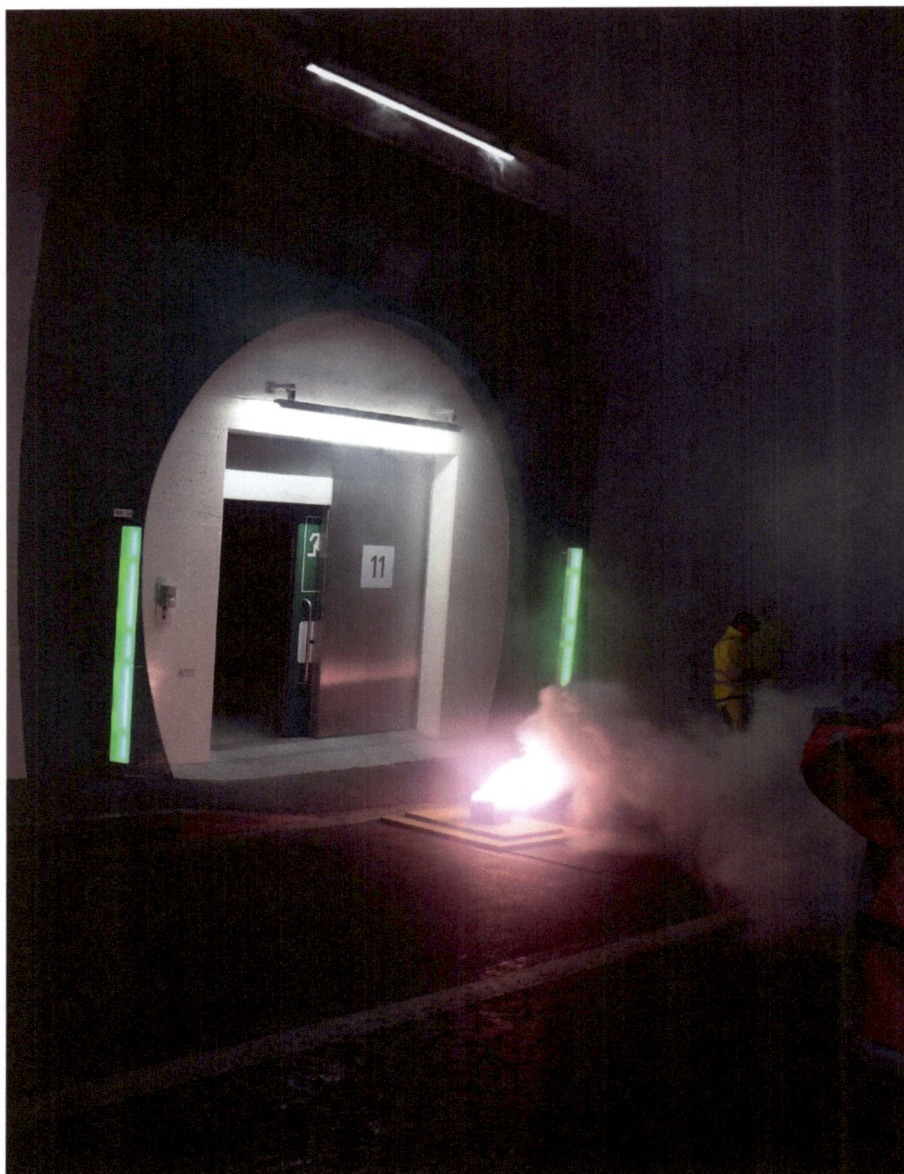

Fig. 40   Test proving the non-entrance of smoke into an open emergency exit door

7 Design hints

7.1 How to proceed

Clear, verifiable goals must be defined. To achieve a goal, different variants of concepts and technical solutions may be proposed. Each variant has its advantages and drawbacks, an optimal range of application, limits of adaptability, uncertainties and costs, depending on the point of view, personal incentives and momentary framing conditions. To find the best solution, 'best' must be clearly defined, and not be subject to vested interests of the decision makers.

What counts is not the design on paper, but the operation in practice. Concepts determine possible operating states to achieve the ventilation goals. The simplest concept that achieve the goals for normal operation and fire ventilation should be chosen. To evaluate where the limits and the benefits of a different ventilation concepts would be, a cost-benefit analysis may be worked out. Useful road tunnel ventilation systems are, with increased complexity and costs:

1. Natural ventilation / no equipment
2. Continuous longitudinal ventilation with jet fans
3. Continuous longitudinal ventilation with jet fans and air exchange stations (Cat. A, B)
4. Concentrated extraction with exhaust duct and controllable dampers (Cat. B, C)
    + jet fans for flow control
5. Transversal ventilation (Cat. C)

Any constructional provisions lead to significant investment costs, in a magnitude that is much larger than the costs of the ventilation equipment. Furthermore, the production of concrete requires a significant amount of energy and emission of greenhouse gases. In this aspect, the longitudinal ventilation with jet fans does not require additional constructional measures, like ducts and buildings, and should be preferred whenever possible[13].

The required operating states must be achieved under various boundary conditions and with different initial conditions, defining possibly various design cases. Most importantly, the airflow induced by the moving vehicles must be calculated and taken into account too.

The design process focuses on:

-   Longitudinal airflow in the tunnel tube
    for different design cases, defined by initial and boundary conditions
-   For more complex ventilation systems, additional exhaust and supply volume flows

and results in specifications of:

-   Necessary ventilation equipment, and interfaces to structural and electrical works
-   Automatic control routines / manual operating procedures

---

[13] Even the longest road tunnel in the world, the 24 km single tube Laerdal tunnel in Norway, has only a continuous longitudinal ventilation.

The design is worked out by simple steady-state calculations, based on experience and simple rules of thumb, for a defined set of critical boundary conditions. The distinction between design and operation is essential. A design case is not necessarily an operating state. Design cases and defined operating states are derived from preconceived scenarios.

Dynamic simulations may be applied for design verification under different initial and boundary conditions. For design and operation, it is to be considered:

- How does the system behave under other boundary conditions than those defined in the design case?
- How robust is the system against the unforeseen?

In reality, scenarios may emerge of which no one had thought about.

As a general engineering rule, uncertainties of the design must be reckoned and safety margins considered. This inevitably leads either to over-dimensioning or to the risk that the goals cannot be achieved. In the first case, you spend more money than necessary. In the second case, the ventilation system does not fulfil its purpose, and the money has been wasted for nothing.

Each specification and requirement should be testable, not only by simulations, but also as realistically as possible on the installed system. While defining the design and operational specifications, the test concept must be considered simultaneously. The design on paper, the test scenarios in the tunnel and the reality differ from one another, especially in case of fire ventilation, which is decisive for the design of most tunnel ventilation systems.

Tunnel ventilation design calculations have been described e.g. by Haerter [54]. All aerodynamic calculation models and parameters for the ventilation design should be regularly validated by measurements which were carried out on a completed system in realized tunnels on commissioning.

As explained in chapter 4, normal operation is not a design case for most tunnels in countries with vehicle emission restrictions, with the exception of long tunnels with high traffic loads. In most tunnels the fire ventilation defines the design case, automatically satisfying requirements for normal operation.

In all tunnels, regardless of the ventilation system, the longitudinal ventilation in the tunnel tube has to be designed, with or without exhaust, for the case of fire. For a simple steady state[14] calculation, the fire ventilation design case is to achieve the so-called critical velocity or an estimated fixed value against adverse pressures, see Fig. 41. The critical velocity can be calculated e.g. using the formula in the NFPA standard [26].

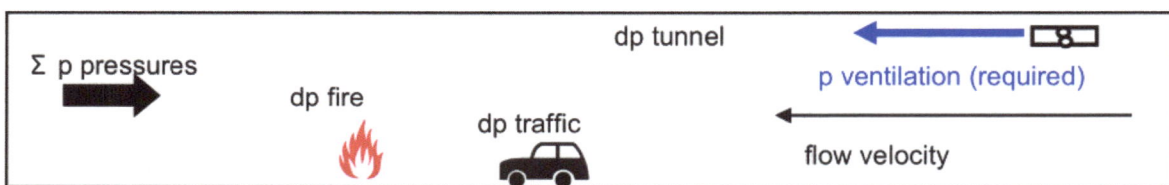

Fig. 41   Model for the design of a longitudinal ventilation

---

[14] In fact, there is no 'steady state' since everything is dynamic. It depends on time framing.

The critical velocity is not a recommended operational state, with the exception of portal zones. Nevertheless, the author suggests to consider it as a basic design case because:

- For short and medium length tunnels, with a longitudinal ventilation system designed to achieve the critical velocity, the dynamic achievement of the required flow velocity according to Table 1 / chapter 5.5 from different initial states is usually ensured.
- The critical velocity may be an operational state in portal zones where flow reversal is strived for according to chapter 5.8.
- The critical velocity is specified by many standards and guidelines, mostly based on request by the fire department.
  'Render unto Caesar that which is Caesar's'

The design usually refers to a defined 'cold' air density. Fire models involve high temperatures which cannot be realistically simulated in the tunnel. Therefore, it is advisable to convert the fire ventilation design conditions to a 'cold state' which defines the test scenarios in the tunnel. The following fire effects are to be considered through corresponding factors:

- Increase in volume due to the combustion of solid matter
- Increase in volume due to the heating up of air
- Obstruction of tunnel cross sectional area due to the fire plume
- Buoyancy due to chimney effect / changes of density

These fire effects should in practice be considered through simple rules of thumb and increased safety factors for the testable design, which must be validated based on real fire incidents or tests, as listed in the bibliography.

The design results in a required number of jet fans, and / or required exhaust and supply flows for more complex ventilation systems, leading to specifications of fans, dampers, nozzles, ducts and auxiliary equipment. Jet fan thrust reduction as a result of non-ideal placement due to obstruction, friction, mutual interferences, etc. is to be taken into account.

Most standards require a minimum redundancy by taking into account the failure of one jet fan group. In practice, multiple jet fan groups may be blocked by the control, because jet fans in the vicinity of the fire location must not be operated. The design must take into account this additional required redundancy.

Besides the steady-state calculation, a dynamic simulation of flow control for different scenarios as described in chapter 7.6 is useful for validation. For longer tunnels, it is essential.

Elaborate three-dimensional CFD simulations are not required for the design of a tunnel ventilation system, but may be useful for design validation in case of complex geometries, as well as for demonstration of smoke stratification for fire ventilation and unexplained physical environment conditions.

7.2 KISS, RAMS, LCC

Those abbreviations are common in the engineering world, but what do they really mean?

KISS

As per the motto 'KISS' (Keep It Simple...), tunnel ventilation concepts and systems should be kept as simple as possible as long as they fulfil the goals and requirements.

Alfred Holt (British engineer and merchant, 1829 – 1911) wrote in 1877:

'It is found that anything that can go wrong … generally does go wrong sooner or later, so it is not to be wondered that owners prefer the safe to the scientific … Sufficient stress can hardly be laid on the advantages of simplicity. The human factor cannot be safely neglected in planning machinery.'

With the words of Edward A. Murphy, Jr. (U.S. aerospace engineer, 1918 – 1990)

'Whatever can go wrong will go wrong'

To err is human and errors always come up in complex technical systems. Risks of failure increase if the system has not been implemented and tested with adequate efforts. The more complex a system is, the higher is the likelihood of error. Accordingly, the efforts rise disproportionately to achieve adequate reliability.

Over a certain level of complexity, control, testing and maintenance of single systems is not economically feasible. That is only viable for mass products, e.g. motor vehicles or electronic devices where adequate testing, respectively practical experience, is automatically ensured by the number of devices and continuous development.

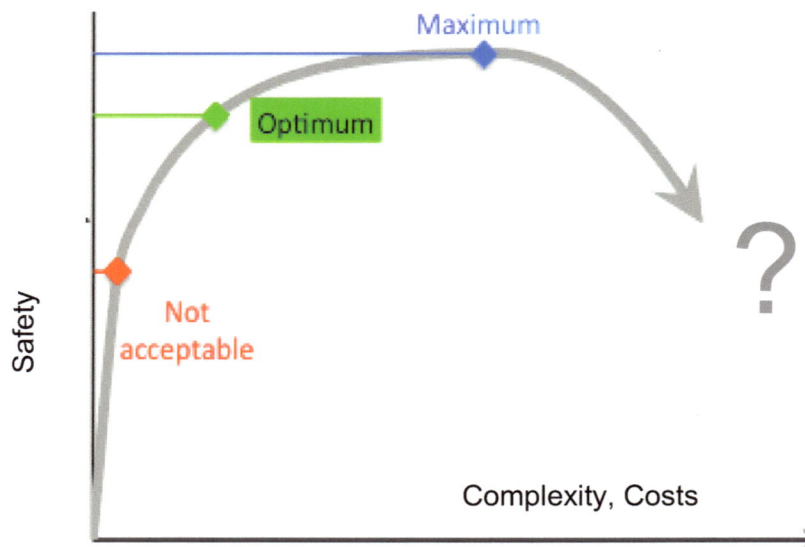

Fig. 42   Safety vs. Complexity / Costs

RAMS

Aspects of Reliability, Availability, Maintainability and Safety (RAMS) must be considered. Any safety system, especially the ventilation, can even exacerbate the situation and increase the damage, when not operated appropriately. Taking into account possible failure scenarios, appropriate fallback modes and redundancy requirements should be defined. It is important to distinguish between systemic and coincidental failures:

- Non-availability of devices and systems during maintenance and construction works
- Failure without operational requirement
- Failure on Demand in case of normal operation
- Failure on Demand in case of an incident

For these failure scenarios, it needs to be determined which consequences do they have on the fulfilment of goals and which measures need to be taken to ensure the required level of safety. In case of ventilation system failures, the following measures need to be considered:

- Risk reduction by limiting traffic and/or speed
- Provision of temporary systems
- Increased readiness of operation staff and emergency forces
- Provision of redundant systems

Unforeseeable failures with operational requirements in case of an incident can in principle be compensated only through system redundancy. The redundancy however significantly increases the complexity and hence also the cost, time and effort for the appropriate measures to ensure complete functionality of the system. In practice, redundancy requirements can in fact lead to an increase in failure probability.

The reliability and failure probability of mass production electronic components is usually known. However, the reliability of ventilation devices, which are produced in small series, can only be roughly guessed. Fans generally are very reliable if they were carefully commissioned and tested and are protected from harmful environment. In practice, most frequent causes of failures are the power supply or the switching devices.

An important criterion for the judgement of safety is that a control system can fail, and its incorrect operation is possible. Therefore, safe states need to be defined for all systems. Similarly, the design of constructional and operational provisions should ensure that even in critical states, in which the foreseen function is not guaranteed, at least no failure with catastrophic consequences occurs.

System error rates should be minimized and contractually guaranteed as long as it is economically justifiable. Sensitive system components should preferably be placed in rooms which are protected from environmental influence and humidity, and accessible for installation and maintenance works.

As a basic condition for maintainability, the corrosive environmental conditions and reduced accessibility in the tunnel tubes are to be considered. Therefore, simple and robust devices with low maintenance requirement and high corrosion resistance should be installed in road tunnels.

For the definition of the maintenance concept, the maximum feasible functional ability of the ventilation system during maintenance must be ensured. Maintenance intervals of the ventilation system and its auxiliary devices should be possibly large.

Another important aspect to be considered is that during the lifetime of a tunnel, the ventilation system would have to be refurbished various times. The lifetime of control systems and instrumentation is even shorter. Refurbishment should be possible under minimal operational obstructions.

LCC

Since in the real world funds are limited, a serious cost benefit analysis should be the basis of any substantial decision, taking into account the life cycle costs (LCC). The lifetime of the tunnel structure may be 100 years or more[15]. Some structural elements, for instance the road surface or the slab in the tunnel ceiling to provide an exhaust duct, must be refurbished beforehand. The lifetime of for electromechanical equipment it is much shorter, and thus must be renewed multiple times during the tunnel lifetime.

The following efforts contribute to the LCC:

- initial investment
  (planning, construction, commissioning and testing)
- investments in later refurbishment works
- capital costs
- insurance
- operation
- energy consumption
- inspection and maintenance
- repair works
- secondary costs in case of non-availability e.g. lost tolls when a tunnel must be closed in case of incidents / for repair works
- compensation for damages in case of incidents

---

[15] The oldest rail tunnel dates back to 1793, but is not in operation anymore. Some ancient tunnels have been in use for more than 2000 years.

7.3   Recommended concepts and fire ventilation goals

Category	Ventilation concept	Fire ventilation goal
0 Short tunnel and/or little traffic	No mechanical ventilation system	
A Unhindered unidirectional traffic	Continuous longitudinal ventilation When required for normal operation / air quality (Very long tunnel, heavy traffic / high emissions): Air exchange stations	Longitudinal flow control for target velocity approx. 1.5 – 2 m/s in traffic direction
B Unidirectional traffic with possibly blocked exit	Continuous longitudinal ventilation Controllable concentrated extraction in case of long urban tunnels with high probability of congestion When required for normal operation / air quality: Air exchange stations	Longitudinal flow control for target velocity approx. 1 – 1.5 m/s in traffic direction or Concentrated extraction with flow control for target velocity approx. 1.5 – 2 m/s from both sides towards the exhaust point
A,B non-incident tubes	-	Build up overpressure + Quick flow reversal + Flow against traffic direction (> 0.5 m/s, fixed setting sufficient)
C Bidirectional traffic	Continuous longitudinal ventilation Long tunnel, heavy traffic: Controllable concentrated extraction When required for normal operation / air quality: Transversal ventilation	Longitudinal flow control for target velocity approx. 1 – 1.5 m/s in measured flow direction at detection or Concentrated extraction with flow control for target velocity approx. 1.5 – 2 m/s from both sides towards the exhaust point
All tunnels, Portal zones, in case of required flow reversal	-	1. Quick flow reversal 2. High flow velocity outward (approx. $v_{critical}$)

Table 1   Recommended ventilation concepts and fire ventilation goals

These recommendations, in particular the flow velocities, represent the actual best practice, and are in accordance with the actual Austrian [28] and Slovak [34] guidelines.

The distinction between different target flow velocities is derived from the assumption that the probability of people being stuck downstream of the fire is lower, but not negligible, for highway tunnels (category A) than for cat. B and C tunnels.

In category A tunnels, smoke removal in traffic direction has the highest priority. People who are situated upstream are safe. Thus, in some countries, the fire ventilation goal in category A tunnels is the critical velocity, or even higher. Unfortunately, such high flow velocities might endanger people downstream of the fire, and are therefore not recommended by the author. Even assuming that in category A tunnels vehicles downstream of the fire location leave the tunnel, the presence of endangered people cannot be ruled out there. However, people situated downstream should be able to escape through the cross passages to the other tube.

In contrast, smoke propagation in category B and C tunnels should be slower than the escape velocity of people, which is assumed in the range of 1.3 m/s [39] to 2.4 m/s [33]. For that, it must be taken into account that the smoke moves faster than the cold airflow.

7.4 Exhaust system and transversal ventilation design

For concentrated extraction, the fans, dampers and ducts are to be designed as follows for fire ventilation purposes:

- The exhaust system should ideally be designed for one open exhaust damper.

- Exhaust quantity at extraction point
  = defined airflow velocity (m/s) x tunnel cross-section (m^2) + safety margin
  or assumed smoke quantity (m^3/s) + safety margin

  With increasing exhaust quantities, the closed loop control of longitudinal flow will become less demanding. On the other hand, the power demand and the investment costs for the construction works would increase.

- The size of exhaust dampers is to be dimensioned for a defined average air velocity in the free open area of the damper.

- The exhaust quantity has to be achieved under the consideration of redundancy requirements. For instance, when taking the failure of one exhaust fan with 100% redundancy as a basis, the fans are designed for the operation of one fan with nominal rotational speed or for the operation of two fans with reduced rotational speed.

- For defining the distances between the exhaust dampers and to the portals, a compromise has to be found between costs and benefits. Usually the structural block partition is to be considered.

- For the design of fans, the leakage values of closed dampers as well as of air ducts must be considered. In exhaust ducts which are separated from the tunnel tube by a slab, the duct leakage is generally much higher that the damper leakage. Moreover, increasing leakage with aging of the tunnel structure is to be considered.

- The air ducts are to be designed for defined flow velocities and permitted pressure. At a specified volume flow, the design results in a compromise between power demand and energy consumption of fans, high pressure loads and leakages in case of high velocities in the duct and high construction costs in case of low velocities.

In transversal or push-pull ventilation systems, the air supply fans are also to be designed. For that, especially the pressure surge in the tunnel tube due to the moving traffic needs to be considered.

7.5 Boundary conditions

For the design of a ventilation system, decisive values for different boundary conditions must be defined, providing a range of possible scenarios that is covered by the design[16]. Important boundary conditions which have an impact on the airflow in the tunnel are:

- Traffic (moving or standing)
- Buoyancy forces (downwards or upwards) due to differences between inside and outside temperatures
- Wind pressure on the portals
- Barometric pressure difference between the portals
- In case of fire: buoyancy due to heat of fire

This leads to the sum of all pressures Σp in Fig. 41 acting on the air in the tunnel. For a simple tunnel with two portals, the pressures can act in two directions respectively, whereupon the target velocity for flow control must be achieved against the determining pressures. In contrast, for the dynamic achievement of a moderate flow velocity by slowing down a strong flow, the pressures act in direction of the airflow.

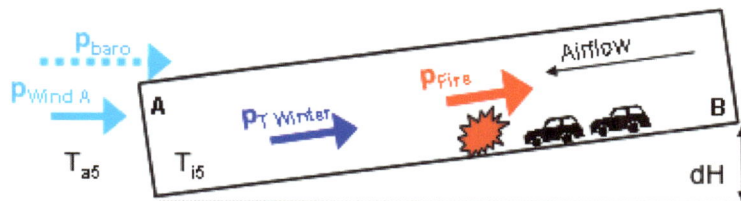

Fig. 43   Determining pressures against airflow in the direction of portal A

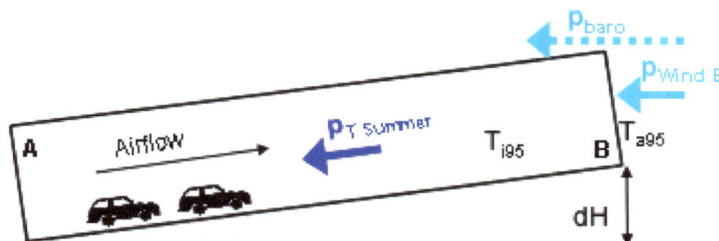

Fig. 44   Determining pressures against airflow in the direction of portal B

---

[16] There is never only one design case. The term 'worst case', which is often used, is nonsense, because there could always be a worse case.

7.6   Dynamic analysis

The desired flow conditions as specified e.g. in Table 1 must be achieved dynamically from any initial state within different boundary conditions. The time to achieve the desired flow condition is essential[17]. An important factor to be taken into account for the flow control in long tunnels is the moving traffic in the first minutes after the fire alarm, dragging the airflow in the tunnel.

The dynamic achievement of desired flow conditions from different initial conditions can be assessed either by using dynamic mathematical models or by adding appropriate safety margins to the stationary design, which must be derived from experience. Finally, it must be tested on site in the tunnel.

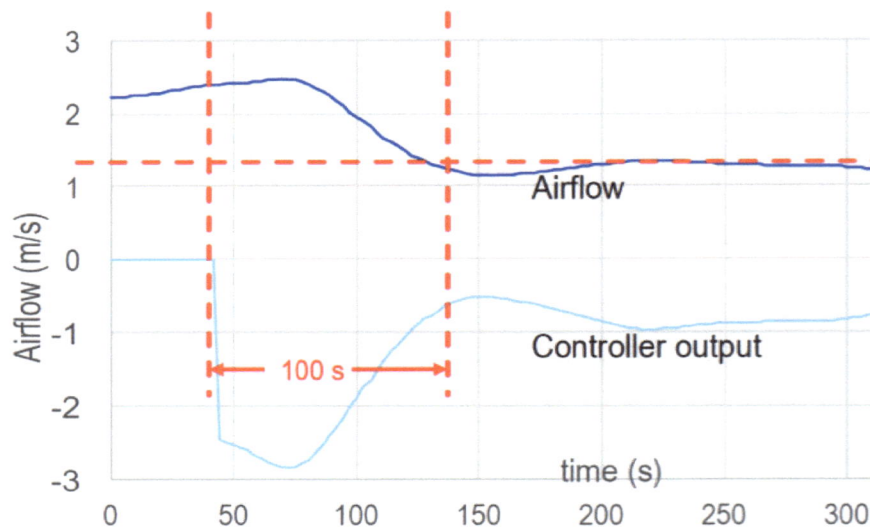

Fig. 45   Example of measured airflow and controller output during fire ventilation tests (target flow 1,25 m/s)

The air can be considered as incompressible for the design. However, pressure surges in the tunnel due to moving traffic must be considered for the design of fans and as an environmental condition. Pressure surges in case of sudden closure of shut-off valves in pipes with a flowing medium are commonly known in the hydraulics field since the 18th century. In tunnel ventilation systems, such surges can occur during following conditions:

-      In exhaust ducts during closure of dampers. Therefore, the dampers and their actuators have to be designed in such a way that no sudden closure is possible.
-      In escape routes with operating overpressure ventilation and closure of open doors.
       This leads to a situation in which e.g. wing doors cannot be opened again for a while after their closure. As a remedial measure, corresponding generously dimensioned pressure release dampers are to be foreseen parallel to the escape doors, that is, to the tunnel tube. In case of sliding doors, this problem is less critical.

---

[17] E.g. max. 2 Min. after the fire detection according to the Slovak guideline [34], or 5 Min. according to the Austrian guideline [28].

8   Technical requirements

8.1   Overview

Overview of equipment and structures that may be required to achieve the ventilation goals:

Ventilation equipment

- Jet fans in the tunnel
- Central fans (Supply, Exhaust) in fan plants
- Dampers
- Auxiliary constructions (e.g. nozzles, diffusers, ducts, guide vanes, support constructions)

Instruments

- Fan supervision (e.g. vibration monitoring)
- Damper supervision (e.g. torque switch)
- Anemometers in the tunnel
- Smoke detectors / visibility measurement in the tunnel
- Gas sensors (CO, NOx) in the tunnel
- Temperature and moisture sensors in the tunnel and outside

Electrical interfaces

- Power supply cables
- Signal cables
- Control cabinets and switchboards
- Variable speed drives (VSD) for fans
- Power supply
- Local control with PLCs
- Fire alarm system
- SCADA
- Communication

Structural interfaces

- The tunnel structure itself
- Air supply and exhaust ducts
  (unlike in building HVAC, ducts in tunnel ventilation systems are usually part of the civil structure, built of concrete)
- Fan buildings
- Impulse nozzles
- Fresh air injection pipes with throttling elements (for transversal ventilation systems)
- Doors

8.2 Quality

The required quality and operational safety of the ventilation system must be based on a defined lifetime, see chapter 7.2. In many tunnel ventilation systems, the fans have been working flawlessly for over 50 years. In contrast, measuring devices, control systems and Variable Speed Drives may be obsolete after a decade. The lifetime of all devices is usually longer than the warranty periods, beside the wear and tear materials.

The environmental conditions in the tunnel can be highly corrosive depending on the climate and traffic load. Strict requirements for the corrosion protection of devices must be applied.

Fig. 46   Highly corroded and damaged jet fan

Stainless steel is generally not more resistant in road tunnel atmosphere than mild construction steel with a high-quality coating, but it may be prone to pitting and stress corrosion cracking, which is not easily visible. High-alloyed, highly corrosion resistant materials are very expensive and not available for all components, e.g. electric motors.

Unfortunately, there are no useful non-destructive test procedures for the quality of corrosion protection. In practice, corrosion resistance can be ensured only when the supplier provides a long warranty period for corrosion protection. This in turn, will require corresponding guarantee conditions with regards to inspection and maintenance.

Along with corrosive and humid atmosphere, strong airflow in the tunnel and pressure surges due to moving vehicles and the strain during cleaning works are to be considered.

8.3   Jet fans

Jet fans and injectors act through momentum transfer from the air jet to the tunnel air. For most tunnel applications, reversible jet fans should be used, with approximately the same thrust in both blowing directions. In special cases, unidirectional jet fans may suffice, which have a better efficiency in the main blowing direction, but achieve only approx. 30% of the nominal thrust in reverse direction.

For the placement of jet fans and injectors in the tunnel, the minimum distances for the momentum transfer, which are approximately equivalent to the length of the air jet, are to be adhered to. Depending on fan size, exhaust velocity and spatial conditions, these minimum distances are usually in the range of 50 m to 80 m. Moreover, the pressure conditions resulting from the escape route ventilation requirements (see Fig. 39) are to be considered. In addition, considerations with regards to failure probabilities, as well as cable lengths are decisive.

Fig. 47   Jet fan group in a tunnel

For twin-tube tunnels with unidirectional traffic, the jet fans and the injectors can be placed on the portals, with blowing direction towards the interior of the tunnel. This guarantees an over-pressure at the cross-passages which are located near the portals. To ensure that no smoke from the other tube can be sucked into the fans, a certain distance is to be maintained between the portal and the fans (see Fig. 56).

8.4 Impulse nozzles

Impulse nozzles act similar to jet fans but due to the supplied air, an additional pressure is built up. The efficiency is worse than that of jet fans. Significant investment costs arise for the construction of ventilation buildings with air supply fans.

An advantage of impulse nozzles is easy access for installation and maintenance works. Moreover, the fans are not exposed to the corrosive tunnel atmosphere. The application of impulse nozzles may be useful when:

- Existing ventilation buildings can be used for refurbishments
- Jet fans cannot be placed inside the tunnel tube due to spatial limitations
- Tunnel closures for maintenance works are extremely restricted
- Tunnels may be temporarily flooded[18]

Fig. 48   Saccardo nozzle in a tunnel ceiling

---

[18] For instance the Stormwater Management And Road Tunnel in Kuala Lumpur, Malaysia

8.5   Exhaust and air supply fans

For exhaust systems, transversal ventilation and impulse nozzles in road tunnels as well as for the overpressure ventilation of escape routes, unidirectional axial fans are common[19]. Centrifugal fans were historically used, but from today's point of view are not suitable for tunnel applications due to their large structure and spatial requirement, and slow controllability.

The exhaust and supply fans generally have a fixed blade angle which can be adjusted during standstill. A variable pitch control may be required in extraordinary cases, but leads to higher investment and maintenance costs. A variable pitch control is useful when the efficiency should be optimized at different operating points, e.g. in road tunnels with transversal ventilation systems which are frequently in operation. Variable pitch control may also be required for the start-up of fans in parallel mode, especially to account for design uncertainties.

Fig. 49   Tunnel air supply fan, with pitot tubes in the inlet nozzle for flow measurement

---

[19] Reversible supply-/exhaust fans were used until the 1990[ies] in semitransversal/transversal ventilation systems, being still common in metro ventilation 'push-pull' systems.

For the start-up and adjustment of the required volume flows, Variable Speed Drives (VSD) are common. Fan, motor, converter, filter and cables must be compatible to each other.

For axial fans, it is important to ensure an undisturbed intake in the direction of fan axis. In case of fans operating in parallel mode, the fan behavior during start-up and operation in single and parallel mode as well as the start-up of a fan during operation of another fan is to be considered. Unstable operating conditions must never arise.

For the design, a reserve from the calculated nominal operating point up to maximum possible pressure (stall limit) with regards to the nominal pressure is to be foreseen, to ensure a safe start-up and operation under all conditions. An excessive design margin would however lead to larger fans and therefore higher construction costs for the ventilation buildings.

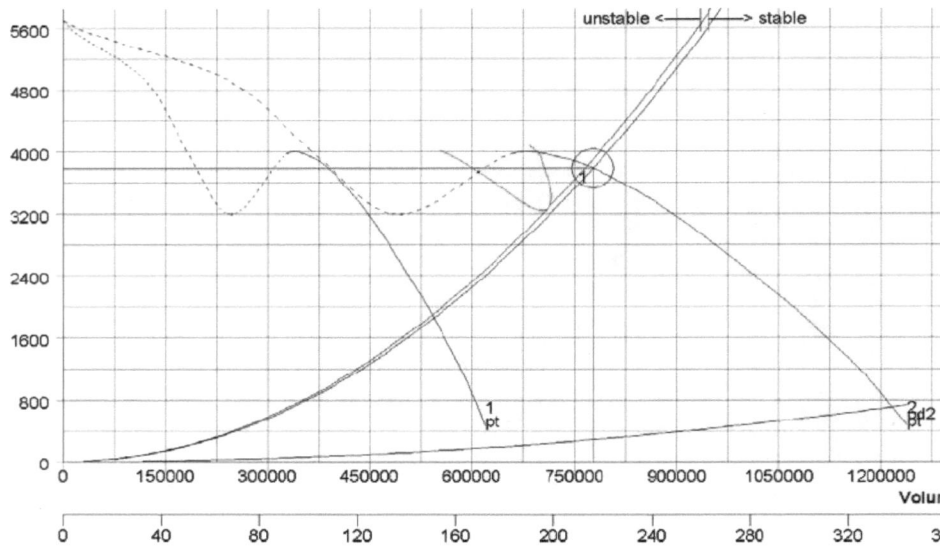

Fig. 50   Characteristic curves of two fans in parallel operation along with system curve (copyright Witt & Sohn AG)

In case of fans that are used during normal operation, pressure surges in the tunnel due to the traffic are also to be considered.

8.6   Silencers

In normal operation, legal noise restrictions must be met. In emergency operation, the noise level must not impede self-rescue of tunnel users and emergency service operation.

Inline silencers are integral parts of jet fans. Large exhaust and air supply fans are usually equipped with splitter attenuators.

8.7 Dampers

For extraction, duct separation and fan closure, generally vane dampers are used. In case of horizontal extraction dampers in a tunnel, the vane axes are usually placed perpendicular to the tunnel axis, whereas in case of vertical exhaust dampers, the blade axes are to be placed horizontally.

A tight power transmission must be ensured between the damper elements, the gearbox and the drive. The damper must not get stuck or blocked in any position whatsoever, nor get misaligned due to external forces acting on the vanes. To avoid a sudden pressure surge in the air ducts when closing the dampers, the closure is to be delayed appropriately.

Fig. 51   Smoke exhaust damper

In case of extraction dampers for which an intermediate position is defined, particularly for line exhaust for transversal ventilation, the opening angle must be set with high precision.

All sealing surfaces between the damper frames and the sub-construction must be sealed such that the requirements related to tightness of dampers are fulfilled when installed on the construction. The sealing material must be airtight and durable throughout the lifetime of the tunnel under the adverse conditions of the tunnel environment.

8.8 Power supply, switchgear and VSD

The power supply is most important for system safety, particularly the ventilation system. In practice, fan or complete ventilation system breakdowns are caused mainly by power supply failures.

The reliability of the entire supply chain (grid – mains connections – transformers – cables – switchgear – Variable Speed Drive / VSD – fan) must be taken into account. Usually the power supply of safety systems should be provided by two independent grids.

Within the scope of available power and aerodynamic design, a quick fan start-up should be strived for. VSD supplying jet fans for closed loop control must also be able to stop the fans quickly. Regenerative converters are suitable for this purpose. Otherwise, braking resistors must be foreseen. Attention should be paid on the heat dissipation due to losses.

Motors which are operated by VSD must have electrically insulated bearings to avoid stray currents and high insulation resistance. When this is not possible, for instance for motors with temperature resistant windings, provision should be made for appropriate filters to limit peak voltages. The design of filters depends on the cable length. Filters in turn lead to an additional voltage drop. Motor, VSD, filters, cables and switching devices must be compatible with each other. To comply with EMC requirements, attention needs to be paid to the limitation of harmonics on the grid side too.

8.9   Temperature resistance

Temperature resistance requirements in standards for smoke and heat exhaust systems were developed in first order for building applications. Exhaust fans for smoke extraction are generally certified in accordance with those relevant standards, although in tunnels, the extracted smoke and air mixture hardly ever reaches high temperatures due to the dilution with cold air and the cooling effect of the concrete walls in the ducts

Dampers for extraction of smoke gases from the tunnel must remain operationally reliable during a certain time period after the outbreak of fire, even in case of high temperatures. Open dampers must not get closed on their own.

The mounting structures of all installations in the tunnel tube, including those of jet fans, must not fail at temperatures at which an intervention by the fire department may still be possible, in order not to put the fire fighters in danger.

In contrast, temperature resistance requirements for jet fan operation are questionable. Jet fans in hot air would not be efficient due to the reduced thrust caused by low air density. Since a jet fan blows the extracted air and smoke back into the tunnel, it is by definition not a smoke exhaust device. Operating a jet fan in a smoke layer would be detrimental and must be strictly prevented. Therefore, failure or blocking of multiple jet fan groups close to the fire zone is considered for the design, increasing the number of installed jet fans.

8.10   Airflow measurement in the tunnel

The information on airspeed and direction in the tunnel is essential for the ventilation control. When no closed loop control of longitudinal flow is required, an anemometer generally suffices per tube and per section for the visualization of flow. For the closed loop control of longitudinal flow, a high precision and reliable airflow measurement is required. Measured values can be falsified through local influences such as obstructions in the tunnel and open doors, or through signal errors.

Therefore, airflow measurements should be based on at least three independent values for plausibility checks. Along with the direction and velocity of airflow, the temperatures are also to be measured to be able to calculate a mass flow balance. In systems with concentrated extraction, the volume and mass flows measured at the exhaust fans might also be included in the calculation of mass flow balance. Leakages of exhaust ducts and closed dampers are to be considered as well.

For the flow measurements in road tunnels, devices measuring in a line across the cross-section by ultrasonic transit-time difference are useful. Point measurements require an increased effort for signal processing.

All anemometers must be calibrated and the raw output should be standardized to a uniform value, by applying calibration factors to consider the local conditions. The calibration can be carried out by a grid measurement in the tunnel, as required by the standard [22].

Fig. 52   Grid measurement in the tunnel for calibration of airflow measurement

Signal processing of measured data must comprise of timely integration, conversion to an equal mass flow along the tunnel / section, correction factors for local conditions, plausibility check of at least three independent values and averaging of plausible values. With the help of a suitable and proven algorithm, values lying outside a defined bandwidth are to be classified as implausible and are not to be used.

The demand for plausibility check leads to measurements at multiple locations distributed over the entire tunnel length. In tunnels with concentrated extraction, the measurements are placed in each portal respectively in each section, each measuring location consisting of three anemometers at short distances.

For positioning, attention should be paid that the anemometers are placed at sufficient distances from the portals, lay-bys and jet fans/ injectors. If an anemometer is located within the range of an air jet from running jet fans or injectors, then its signal should be suppressed.

8.11 Fire detection

Fast and reliable fire detection is a prerequisite for the prompt automatic start-up of fire ventilation as well as of other safety systems. On waiting for the confirmation by an operator, the chances of human error increase and precious reaction time is wasted. Therefore, a high reliability and low false alarm rate of detection devices should be strived for. The following systems are usually applied in road tunnels:

- Linear fire alarm system based on thermal sensors
- Smoke detectors / opacity meters
- Video detection (CCTV)

Knowledge of the exact fire location is essential, since jet fans in the vicinity of the fire must not be operated, and a smoke extraction should be as close to the fire as possible. Fire ventilation based on the wrong fire location could have a detrimental effect. Most important in tunnels with an extraction is that after the first alarm has started the smoke exhaust, any subsequent fire alarms must be ignored. The extraction point may be changed manually, but the fire ventilation must never allow multiple extraction points[20].

For all detection devices, the quicker the detection, the higher is the false alarm rate. Though thermal sensors are widely used, in practice they react either delayed or not at all, especially in case of fires inside a vehicle, in case of smoldering fires with less heat generation and/or at high airspeed.

Video detection is able to detect smoke very quickly, but reliability is not guaranteed, since the false alarm rate is very high. Fire detection by CCTV must be confirmed by an operator, who is again dependent on the information received from cameras in the tunnel.

Conventional smoke detectors used in buildings are not suitable for use in tunnels because of harsh environmental conditions, especially humidity. In practice, opacity meters have proven to be the most reliable devices for smoke detection in road tunnels. If installed at short distances, they not only make quick smoke detection possible, but can also be used for the control of normal operation ventilation. Therefore, it is recommended to equip all tunnels, even those without a ventilation system, with smoke detectors in regular distances. Additional smoke detectors should be installed at each portal, as well as in all fresh air inlet openings for ventilation of technical rooms, escape routes or air supply in case of transversal ventilation.

Such smoke detectors in the form of opacity meters with low resolution and large measuring range are suitable as a primary source of fire incident detection in road tunnels, but they cannot determine the exact fire location. Therefore, a fire alarm system based on linear temperature sensors may be required additionally for exact localization. This must guarantee a determination of fire sections corresponding to at least half of the distances between jet fans and / or exhaust dampers.

---

[20] In the disastrous Gotthard tunnel fire in 2001, the multiple fire alarms led to linear extraction in several sections over more than 8 km, which significantly increased the spread of smoke.

A clear distinction between stationary and mobile sources of smoke is required. Based on the distance between the measuring instruments and the time between exceedance of threshold values at individual devices, the direction and speed of smoke propagation can be calculated. Moving vehicles travel much faster than the airflow. As long as the burning vehicle is moving, an exact location is not possible.

8.12 Opacity and pollutant measurement

Opacity measurement is used to monitor the air quality in the tunnel, providing the guide value for ventilation control.

In most tunnels, where the ventilation is rarely ever in operation, a resolution of around 1 km^{-1} extinction coefficient suffices for normal operation, as is achieved by smoke detectors. For smoke detection, a measuring range of up to around 1000 km^{-1} extinction coefficient is required. In contrast, high precision opacity meters with resolution of around 0.1 km^{-1} extinction coefficient are only useful when the ventilation is expected to operate frequently and therefore needs to be optimized for low energy consumption.

Note that standard smoke detectors as used in buildings are not usable in the tunnel environment. Tunnel smoke detectors are in fact simplified opacity meters.

Additional pollutant measuring instruments for CO and NO_2 can be installed on requirement, based on the calculation of fresh air demand or on basis of requirements for emission control as demanded by Environmental Impact Assessment (EIA). However, in practice, opacity as the only relevant input parameter for normal operation ventilation has proven effective.

For opacity meters near the tunnel portals, it must be ensured that fog is not detected as smoke. Alternatively, fog detectors can be installed near the portals for plausibility checks.

8.13 Device monitoring

A signal indicating whether a fan is working or not can be ascertained e.g. by monitoring the airflow or pressure difference. If frequent fan operation is expected, online vibration and bearing temperature monitoring is recommended. For jet fans that are rarely in operation, offline sensors or regular vibration checks during maintenance are sufficient.

In case of fans that are operated by Variable Speed Drives, the winding temperatures as well as operating status and fault signals of the VSD are to be monitored.

A precise volume flow measurement of exhaust or supply fans may be required if used as a reference value for closed loop control of longitudinal flow in addition to the airflow measurements in the tunnel. Such a volume flow measurement must be calibrated.

For jet fans in the tunnel ceiling, a position switch should detect when the suspension fails and the fan could fall down, as may happen e.g. by a collision with an oversized truck.

For the monitoring of dampers, end and torque switches, as well as operating status and fault signals of damper drives are to be monitored. In case of dampers with an intermediate position as required for transversal ventilation, a continuous position status monitoring is required.

8.14   Control system

The ventilation system with respective sensors should function as highly reliable autonomous system independently of any superordinate control system (SCADA). Where tunnels are controlled by an operator team and not automatically, serious training, education and regular service drills of the operators must be ensured.

Basic functionalities like power monitoring are usually implemented directly on the hardware (Switchgear and VSD). The programmable logical control (PLC) provides the following functionalities:

-   Signal evaluation of airflow measurement, including plausibility check
    between anemometers and determination of mean mass flow for each branch
-   Signal evaluation of smoke detectors, including determination of fire location and
    recognition of moving fires.
-   Signal evaluation of opacity and air quality in the tunnel
-   Equipment monitoring (Fans, VSD, dampers)
-   Functionality of operational requirements for normal operation:
    Switching of ventilation according to opacity / air quality, including hysteresis
-   Functionality of operational requirements for fire ventilation:
    Closed-loop flow control, smoke exhaust close to fire location (if available),
    prioritization and blocking of jet fans according to detected fire location
-   Communication to SCADA, fire alarm system and other systems

In case of control system failure, the possibility of manual operation should be provided.

By integrating the PLC, VSD and switchgear in Control Units close to the fans, increased system redundancy can be achieved, and problems with cabling and EMC issues can be avoided.

Fig. 53   Jet fans with control units containing VSD, PLC and switchgear

8.15 Structural provisions

In case of continuous longitudinal ventilation with jet fans, there are usually no constructional requirements from the ventilation side. Ventilation ducts and central stations for exhaust and air supply fans lead to significant investment costs for constructional provisions. That's why from a cost-benefit point of view, simple continuous longitudinal ventilation systems should be strived for.

In twin-tube tunnels, transfer of polluted tunnel air from the exit portal of one tube to the entry portal of the other tube can occur. Generally, this is a problem only in very long tunnels and can be avoided by shifting of portals in their respective traffic direction. The feasibility of portal shift is however dependent on the terrain and has a high impact on construction costs.

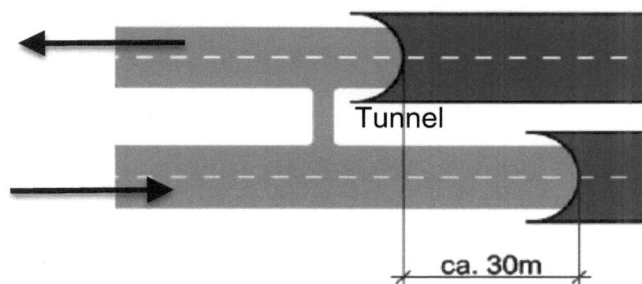

Fig. 54   Required portal shift (from [27])

Separating walls, in comparison, have less influence in this regard. Practical experience shows that especially in case of fire, smoke transfer from one tube to the other cannot be avoided by separating walls. Therefore, it is of utmost importance that smoke transfer is avoided by quick detection of smoke at the portals, closure of the tunnel for traffic and the immediate start of fire ventilation operation in both tubes.

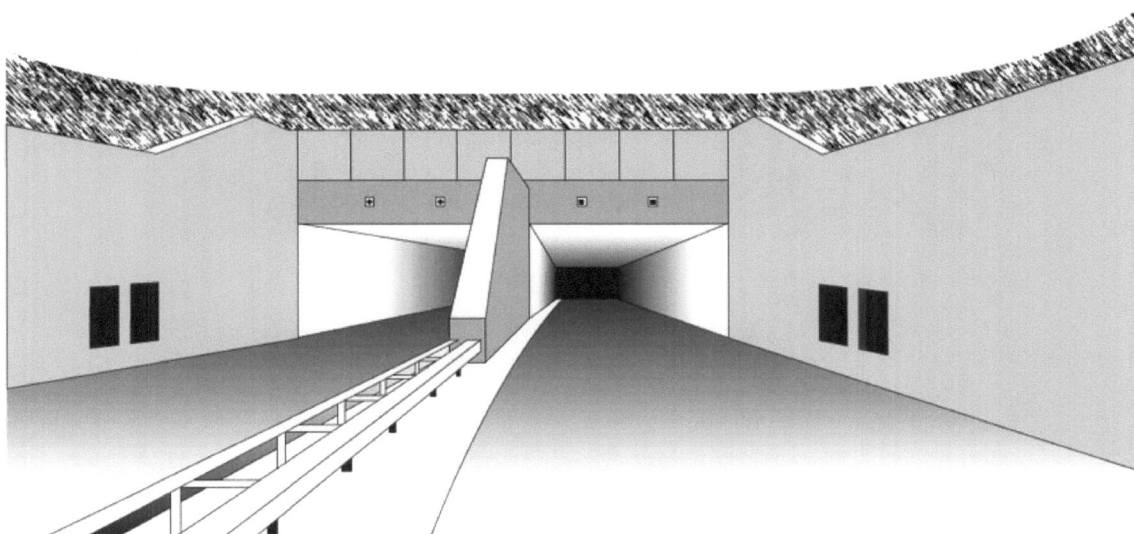

Fig. 55   Theory: Dividing wall (from [6])

Fig. 56   Practice: Smoke spread over a dividing wall during a smoke test

Fig. 57   The solution: Preventing smoke spread by appropriate ventilation (see Fig. 34)

However, separating walls are useful when a quick smoke detection and reaction of the ventilation system cannot be guaranteed, particularly for short tunnels without mechanical ventilation.

For axial fans, attention should be paid towards a uniform and undisturbed intake area in the fan axis. In fan buildings, the requirements for the installation and dismounting of fans as well as maintenance works are to be considered.

Fig. 58   Intake of two parallel exhaust fans in a fan building

Exhaust ventilation shafts should blow out vertically. The cross-section of the shaft should be designed for a minimum exhaust velocity in all states of normal operation and fire ventilation.

For the placement of fresh air inlet openings for the ventilation of technical rooms, of escape galleries or of transversal ventilation systems with air supply, attention must be paid on sufficient distances to the portals or to exhaust shafts to prevent intake of polluted tunnel air or of smoke in case of fire. For that, the local conditions and possible states of airflow in the environment are to be kept in mind.

The structural design of all ventilation ducts should be optimized aerodynamically as long as it is economically feasible, limiting the flow velocity in the ducts, the fan power and the energy consumption during operation. Moreover, the ventilation ducts, as well as fans, dampers, auxiliary equipment and sound attenuators, should be accessible for maintenance.

In case of exhaust ducts for concentrated extraction, tightness is an important factor. Design values of the leakage must be assumed for the exhaust system design. Those design values must be achieved by the constructive design, taking into account that the leakage of air ducts increases with time. The costs for sealing the exhaust ducts are to be compared with the costs for more powerful fans and power supply.

8.16 Testing

Defects in safety systems which are practically never in operation become first visible when an incident occurs, but then a failure can have disastrous consequences. A system is supposed to work reliably only when it is carefully tested. If no faults are found, that may an indication that the testing was not diligent enough. All functions and operating states of the safety systems must be tested as realistically as possible but with justifiable efforts. A lot would go wrong and would require to be corrected. This demands time and costs which rise disproportionately with increasing complexity. This aspect is particularly important for tunnel ventilation systems.

The functionality of single devices and achievement of design criteria is usually proven by Type Test Certificates, Factory Acceptance Tests (FAT) and Site Acceptance Tests (SAT), which are defined in relevant standards.

Important is the proof of functionality and achievement of goals on the complete installed system under different initial and boundary conditions. All functions and operational states are to be tested as realistically as possible, but with justifiable efforts. Therefore, the respective tests should be defined right at the stage of defining the requirements.

The design of the longitudinal ventilation must be recalculated to the airflow in an empty tube without external influences. The longitudinal flow velocities in the tunnel are to be measured in both directions. By evaluation of the airflow measurements, in particular the dynamic step response after switching jet fans on and the decay curve after switching them off, the aerodynamic tunnel parameters can be derived, especially the controller parameters (see Fig. 59)

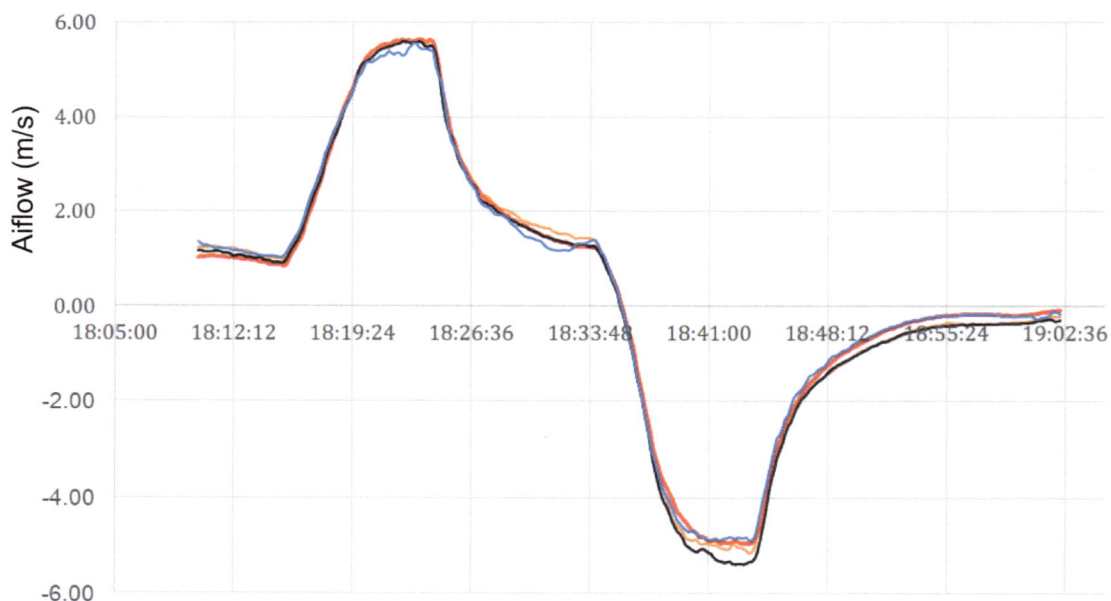

Fig. 59   Airflow measurements with step response and decay curves in both directions

Ideally, the measurement of longitudinal flow takes place simultaneously with the calibration of anemometers in the tunnel. During the measurement of the longitudinal flow, all openings in the tunnel such as doors, dampers, empty conduits, etc. must be closed.

In case of axial fans without pitch control, the blade angle must be adjusted in accordance with the real conditions during commissioning, so that the highest possible volume flow is achieved without exceeding the stall point and reliably ensuring a safe start-up with sufficient reserves under all possible conditions.

In plants with exhaust ducts for concentrated extraction, the leakage of ducts and closed dampers must be measured as well. Duct leakage should be measured on commissioning and preferably after a few years of operation, to assess the effective state.

Fig. 60   Measuring grid for leakage measurements in an exhaust duct

In twin-tube tunnels as well as in overpressure-ventilated escape routes, the forces required for the opening of escape doors and the flow velocity through open escape doors are to be measured during operation of fire ventilation in the tunnel in both tubes and during operation of overpressure ventilation, under simulation of determining external pressures.

Adherence to the noise limits is to be proved by sound measurements.

At the Site Integration Tests (SIT), all programs must be tested in the tunnel for many different scenarios, including random failure scenarios. The function of the ventilation system in collaboration with other safety systems like incident detection, signalization, lighting, etc. is to be assessed.

Simulations of external pressures with the help of mobile jet fans are essential to test the design cases and other defined scenarios.

Fig. 61   Mobile jet fan for simulation of external pressures

During the tests in the tunnel, all measured values, especially the airflow measurements must be continuously recorded. These play a decisive role in assessing the function of the ventilation system. According to the dynamic behavior, the closed loop control can possibly be optimized. Failure scenarios should be tested by performing random sampling of failures of individual equipment, control and communication devices, including possible power supply failures.

After successful completion of all integration tests, fire and smoke tests are recommended for the visualization of fire ventilation for selected scenarios.

The scenarios are defined based on the following parameters:

-       Fire ventilation programs with longitudinal ventilation – with smoke extraction (if applicable)

-       External pressures from one side – from the other side
        ('design scenarios' and / or 'probable scenarios')
        The pressures may be simulated by means of mobile jet fans.

-       Smoke quantity corresponding to a personal car fire or bigger, as feasible with justifiable efforts.

-       Stationary source of smoke – mobile source of smoke on a vehicle.

The test smoke should fulfil the following requirements:

- The physical characteristics of the test smoke, especially the optical density should be similar to that in case of a real vehicle fire.

- The smoke detection must react reliably, even with fog suppression.

- The produced heat must not damage the installations in the tunnel but should induce thermal buoyancy and smoke stratification.

- The test smoke must be harmless with regards to toxicity and possible health hazards.

- The test smoke must not be corrosive and must not leave behind any deposits on the installed equipment in the tunnel, particularly on measuring instruments and cameras.

Fig. 62   Simulating a fire: Hot smoke test in a tunnel with smoke exhaust

For tunnels in operation, the fire ventilation should be automatically tested at regular intervals, each time with different scenarios. This can be carried out at times of less traffic (e.g. at night), when the normal operation ventilation is not running. In the process, the entire measurement data of airflow measurements and fan operation is to be recorded and analyzed. To save energy costs during the tests, exhaust fans may be operated at low rotational speed.

8.17   Service and maintenance

Service and maintenance works on the ventilation system are to be carried out according to the manufacturer's specifications, generally once a year. Special attention needs to be paid to the cleaning and repair of occurring corrosion.

If service and maintenance works are to be performed on measuring instruments and jet fans in the tunnel tube, the tunnel has to be closed. For works on devices outside the tunnel tube, the tunnel does not need to be closed, but the risk resulting from non-availability of the devices needs to be reduced, for example by carrying out these tasks at night during less traffic.

After completion of maintenance works, all devices are to be visually examined and performance checks need to be conducted. Measuring instruments might be recalibrated if necessary and vibrations of operating fans should be measured.

It is recommended to periodically execute the following tests during tunnel closures:

-   Calibration of measuring instruments/ i.e. airflow measurement together with assessment of longitudinal ventilation (see Fig. 52).

-   For extraction or transversal ventilation:
    Calibration of volume flow measuring instruments of supply and exhaust fans and leakage measurements of ducts and dampers.

-   Fan vibration measurements and bearing condition monitoring on fans

-   Measurements of escape route ventilation parameters (door opening, flow through open escape doors).

-   Integral tests and hot smoke tests.

By that, effects resulting from ageing of components and devices can be assessed, and possible refurbishment may be tackled.

Fig. 63   Maintenance works on a jet fan.

9 Fire incidents – case studies

9.1 Mont-Blanc 1999

The Mont-Blanc tunnel is situated below the highest peak of the Alps between France and Italy. At the opening in 1965, it was the longest road tunnel in the world with 11.6 km length. The tunnel was operated by two independent companies, one French and one Italian, from two distinct control centers. Communication and coordination between the two operating companies, in two different languages, was quite poor. On March 24, 1999 a fire started inside the tunnel, originating from the engine of a Belgian truck loaded with margarine and spreading over various other vehicles. Totally 34 vehicles including 20 trucks burned. According to the official report [51] the fire was caused most probably by a cigarette stub in the air filter of the truck engine. The fire site, where the truck had stopped, was situated near the center of the tunnel. Fire detection was delayed, and the number of cars in the tunnel was unknown to the operators. Traffic signals, preventing vehicles from advancing further into the tunnel, were either not obeyed or not working at all.

The smoke extraction capacity of the tunnel ventilation system was under-designed, like in most tunnels at that time. After the fire alarm, the Italian operators blew fresh air into the eastern side of the tunnel, thus providing a safe zone for the people in the Italian part, but accelerating the smoke all over the western French part. There, 27 tunnel users died resting in their vehicles, and 10 died while escaping on foot. Two rescue workers lost their life too. However, many people have saved their lives despite the unfavorable conditions, because they immediately left their vehicles and fled on foot to the remote portals, as did the driver of the truck that started the disaster.

Fig. 64   Inside the Mt. Blanc tunnel after the 1999 fire (public domain)

An important aspect is that the Mont-Blanc tunnel was equipped with shelters every 600 m, supplied with fresh air, but without access to the open. In a shelter close to the fire zone, two people died. The shelters had a two-hour fire protection rating, while the fire exceeded temperatures of 1000°C and lasted for 53 hours. It took five days for the fire site to cool down sufficiently to enable unprotected access. As a consequence, subsequent new tunnel safety rules banned closed shelters, demanding escape ways to the open. This requirement ignored the fact that some tunnel users and fire fighters have survived the Mont-Blanc tunnel fire only thanks to those shelters, where they were protected from the deadly fumes, until they were rescued by the fire department. Also, the importance of a safe drainage system was derived, since the spreading of burning liquids over the road surface had caused further fire propagation.

9.2 Tauern 1999

Only two months after the Mont Blanc tunnel fire disaster, 12 people died on May 29, 1999 in a rear-end collision and a subsequent fire in the 6.4 km long Tauern tunnel in Austria. A sleep deprived truck driver crashed into a column of several vehicles, including a heavy goods vehicle with lacquer tins. That column had built up in front of a stop light at a construction site inside the tunnel. Eight fatalities resulted from that accident, regardless of the following fire, where 20 cars and 14 trucks burned down during 14 hours. Temperatures in the tunnel exceeded 1000°C. Three victims were found dead in a vehicle that was 100 m away from the fire site towards the portal, whereas people that were closer to the fire had successfully escaped. The last victim was found 800 m from the fire towards the interior of the tunnel. He was suffocated when leaving an emergency call niche. Three people took refuge in another emergency call niche and were saved later by the fire rescue service.

Fig. 65   Fire in the Tauern tunnel 1999 (public domain)

The drivers and operators in the Tauern tunnel had been sensitized to tunnel fires, because the Mont-Blanc disaster was present in all media by then. Approximately 80 people managed to escape from the tunnel. The ventilation had pushed the smoke more than one kilometer towards the inside of the tunnel. Later, the flow direction was changed on demand of the fire department, to save the people in the emergency niche.

9.3 Gotthard Road Tunnel 2001

The 16.9 km long Gotthard road tunnel in the Swiss Alps was the world's longest road tunnel at that time, being situated on an important European North-South traffic link. On October 24, 2001, a frontal impact between two trucks occurred, caused by a Turkish driver in a Belgian truck, resulting in a fire in which 23 vehicles were totally burned or damaged. The fire spread so quickly that it could not have been fought by hand-held fire extinguishers. Burning tires emitted dense smoke. The driver that caused the crash was found dead 300 m away from his vehicle. According to the investigations, he had been drunk.

Fig. 66   Fire in the Gotthard road tunnel 2001 (public domain)

When the accident occurred, the refurbishment of the tunnel ventilation system towards a controllable concentrated smoke extraction had already been in progress. Unfortunately, the old system with line exhaust was still in operation when the incident happened. The fire detection system gave further alarms, which activated the linear smoke extraction system in other sections, increasing the air velocity at the fire site and further fanning the flames and pulling the smoke towards the inside of the tunnel[21]. The picture in Fig. 66 was taken from the safe side, but behind the fire, the tunnel was smoked up over a length of approximately 2.5 km.

The Gotthard road tunnel is equipped with emergency exits to shelters in distances of 250 m, which are connected to a parallel escape tunnel. Those escape facilities saved at least 30 tunnel users. Others escaped towards the south portal. Most of the 11 people who died could have survived if they had reacted quickly and properly, fleeing immediately to the shelters. Five victims died staying in their vehicles. One driver was found next to an emergency exit.

---

[21] A basic rule in tunnel fire safety issues is that only the first confirmed alarm must release the automatic response. Working on the ventilation design and control, the author has pointed that out in a technical report several weeks before the fire incident happened.

9.4   Viamala 2006

The situation in the 750 m long Viamala tunnel was different from the previously discussed fire incidents. The old, steep and narrow tunnel has tight curves, where the sight distance is limited. On September 16, 2006, an accident between a bus and two cars resulted in the outbreak of a fire near the lower end of the tunnel. The fire grew quickly, and the chimney effect caused uncontrolled spread of smoke and toxic gases to the upper portal. Seven vehicles were trapped in total. Subsequently, the tunnel was filled with smoke over the whole length very quickly, before any safety system could react. The jet fans of the ventilation system stayed switched off, since they were of no use to control the smoke spread effectively.

Fig. 67   Burnt car and bus in the Viamala tunnel 2006 (public domain)

Nine people died, some due to the accident, others were poisoned by the fumes. A family of four died in their car, another four people were found dead on the road surface in the tunnel, and one truck driver died later in the hospital from his injuries. However, 21 tunnel users, including an ice hockey team, were able to escape.

When the fire fighters arrived on site, the tunnel equipment in the vicinity of the fire site had already been destroyed and fallen to the surface. Downstream of the fire, they had to search for people under zero visibility conditions. Finally, the fire could be extinguished and the smoke was removed using a mobile jet fan.

The Viamala tunnel had been identified as dangerous before, but planned safety measures had not been implemented yet by then. Ten years later, a ventilation refurbishment program with fast smoke detection and controlled ventilation with more powerful jet fans has been successfully completed and a parallel escape tunnel had been constructed. Nevertheless, the primary cause of the accident, which is the insufficient sight distance at the allowed speed in the narrow tunnel, has not been addressed.

In 2016, a fire occurred in the escape tunnel construction site, and the smoke passed to the main tunnel through partially open cross passages. Almost the whole tunnel was filled with smoke, impairing the traffic. One driver suffered from light smoke poisoning. All safety systems worked well, but unfortunately the applied mode of operation was not useful for this particular situation. It was an unexpected event that nobody had thought about previously.

9.5 Yanhou Tunnel 2014

One of the worst road tunnel fire disasters was reported in northern China on March 1st 2014. Unlike the other described fire incidents, where the author has obtained personal information, the following information is taken exclusively from internet sources.

Two methanol tanker trucks crashed during traffic congestion inside a short twin bore highway tunnel on the Erguang Expressway. The drivers tried to examine and disentangle the trucks, triggering a fire that quickly spread to other trucks, carrying coal and other flammable materials. About 100 Minutes after the crash, a liquid natural gas tanker exploded. It took 73 hours to extinguish the fire.

Fig. 68   Smoke exiting the portal of the Yanhou tunnel (public domain)

Many important safety measures seem to have been neglected. A queue of coal trucks built up inside the tunnel, and even after the outbreak of the fire, the tunnel was neither closed to traffic nor evacuated. The cross passages could not be used for escape purposes. At least 31 people died, 9 were missing and 42 vehicles were destroyed. But even then, the death toll was less than 5% of the average daily number of fatalities on China's roads.

9.6   Seelisberg 2017

In the 9.25 km long Seelisberg highway tunnel in Switzerland, on 10. January 2017 a driver noticed that her car was on fire. She stopped in the tunnel and left her vehicle. Shortly thereafter, the car was fully burning. Fire fighters approached the fire from both sides, working with breathing equipment. In short time, the fire was extinguished. After the destroyed car had been removed, the tunnel was reopened for traffic.

Besides the burnt car, no personal or material damage had occurred. The safety systems, namely the fire ventilation with smoke exhaust, had worked as foreseen, and no people were in danger.

An analysis of the smoke detectors showed that in the first phase, the tunnel was quickly filled up with smoke, which was diluted by the strong airflow induced by the traffic. After detection, the traffic was stopped and the smoke exhaust started. In the exhaust zone, high smoke concentrations with almost zero visibility were achieved. Since the exhaust was opened downstream, in traffic direction, nobody was endangered.

Fig. 69   Burnt car in the Seelisberg tunnel (copyright Flüelen Fire department)

However, in another, similar fire incident, erroneous fire localization by the smoke detectors had led to an opening of dampers upstream of the burning vehicle, leaving people trapped in the smoke (see Fig. 31). Even then, luckily nobody was seriously harmed.

9.7   San Bernardino 2018

On May 18 2018, a bus occupied by 22 people caught fire while driving through the San Bernardino tunnel, a 6.6 km long bidirectional tunnel in the Swiss Alps in service since 1967. The bus stopped about 500 m from the northern portal. Three other cars got stuck at the fire location. All people could escape by own means through the emergency exits, which lead to a combined fresh air duct and emergency tunnel below the road surface. The emergency exits had been built a decade ago together with a complete refurbishment of the tunnel when the ventilation system was enhanced with a concentrated extraction and flow control by jet fans.

Local fire departments arrived from both tunnel portals to the incident site. While the fire fighting operations were in progress, the heat from the fire led to the melting of cables, in particular the fire detection cable. By that, a second fire location was displayed on the SCADA. The operator in charge got the message 'do you want to change the fire zone?' on his screen. How would you have reacted under stress, without appropriate information? The operator confirmed, and the smoke extraction was changed to the new fire location. Thus, the smoke was spread in the tunnel to the new extraction zone. The fire fighters were suddenly surrounded by smoke. Although they worked with breathing apparatus, that led to a dangerous situation. One fireman, whose breathing apparatus was exhausted, got unconscious, but could be saved by his colleagues.

Fig. 70   Burnt bus in the San Bernardino tunnel (copyright canton Graubünden police department)

9.8   Gleinalm 2018

The 8.3 km long Gleinalm tunnel had been built in the 1970ies as a single tube with bidirectional traffic. As required by the European Safety guideline [7] for many alpine tunnels, a second tube was built to provide unidirectional traffic until 2019. While the construction works were in progress, a serious bus fire had occurred in 2016, without fatalities, but material damage which led to a three-week tunnel closure.

On October 5th 2018, an exceptional transport with a mobile crane on a truck started to burn due to spilled hydraulic oil in the newly built tunnel tube, which was operated with bidirectional traffic while the old tube was refurbished. The truck driver tried to extinguish the fire manually, but failed. Then he continued to drive out of the tunnel, but the burning vehicle finally broke down.

Meanwhile, the fire had been detected by the linear heat detection cable and the fire ventilation with concentrated smoke exhaust started as planned. But because the burning vehicle moved a few hundred meters further, the smoke spread through the tunnel and would have seriously endangered the people in that zone. Luckily, the exceptional transport was accompanied by an escort vehicle which had stopped and prevented further vehicles from driving into the smoke zone. Due to the heat, fire fighters could not approach the fire location immediately, and had to wait until the heat release rate had decreased. Finally, 83 people could be evacuated; three people suffered from smoke poisoning, but no serious injuries, nor fatalities, occurred. The new tunnel had been destroyed in the vicinity of the fire location. However, the repair works could be completed sooner than expected, since the construction companies were already on site.

Fig. 71   Fire fighters cooling down the burnt truck in the Gleinalm tunnel
               (copyright fire department St. Michael)

9.9   Salang 1982

The worst road tunnel fire in history happened in the Salang tunnel during the war in Afghanistan in 1982. A truck exploded, and a column of Soviet military vehicles burnt out, while the tunnel portals were closed for escape. The indicated number of casualties is between 176 and 3000. The Soviets claimed it was an accident, the Afghan Mujahedeen that it was a deliberate attack. The full extent of the tragedy may never be known. War incidents, where deliberate damage is pursued, do not represent realistic scenarios to assess road tunnel safety in peacetime.

Fig. 72   Inside the Salang tunnel (public domain)

9.10   Résumé

Fire incidents happen regularly, but with low frequency. Only in very long tunnels with high traffic, the numbers are statistically relevant. For instance, numbers of fires from projects where the author has been involved are as follows:

Tunnel / category	Length (km)	Average traffic (vehicles per day)	Average number of fires/year
Elbe tunnel, Hamburg / B	3 x 2.6 km + 1 x 3 km	up to 160'000	10 - 12
Gotthard road tunnel / C	16.9 km	17'500	4
Seelisberg tunnel / A	2 x 9.25 km	22'000 (up to 43'000)	1 - 2

Table 2

Most fire incidents don't result in serious damage. Safety systems work as expected, people can escape, and fire fighters can control the fire quickly. The described fire in the Seelisbergtunnel in 2017 is a typical example. Often, only smoking vehicles are observed, not resulting in a fully grown fire. Serious fires are extraordinary events. Occasionally, people suffer from smoke poisoning, and the tunnel structure is affected in the immediate vicinity of the fire location.

Unfortunately, knowledge obtained from past experience and inductive reasoning is not a guarantee that it will be the same in the future. This is known as the problem of induction. In the last century, smoke control in tunnels generally did not work by today's standards. Nevertheless, mostly nothing happened, with a few exceptions[22].

The fire incidents in the Mont-Blanc tunnel (France/Italy) and Tauern tunnel (Austria) in 1999 as well as in the Gotthard Road tunnel (Switzerland) in 2001, led to public media attention and discussions about tunnel safety. Totally 71 people lost their lives in those three incidents, and subsequently the tunnels were closed for months or even years for refurbishment. This led to traffic diversion to less suitable, more dangerous routes. Beside the ecological impact of additional kilometers driven, the increase of accident rates on the alternative routes probably may have led to further casualties as a secondary effect.

A key role in all those disasters played the smoke propagation and tunnel ventilation as well as the availability of safe shelters and escape-ways, demonstrating the usefulness and limits of safety measures. All victims suffocated from toxic gas, except the people who had been killed by the preceding accidents in the Tauern and Viamala tunnels. All four tunnels had one tube with bi-directional traffic, therefore there were vehicles and people on both sides of the fire location. Only the Gotthard tunnel had emergency exits through cross passages to a parallel escape tunnel. In the Mont-Blanc, Tauern and Gotthard fire incidents, the tunnel ventilation was not only unable to control the spread of smoke in the long tunnels, but even accelerated the smoke propagation.

The media impact of the incident in the Viamala tunnel (Switzerland) in 2006 with 9 fatalities was lower. By then, the media hype had faded. The short, steep tunnel was quickly filled with smoke due to the chimney effect, and the tunnel ventilation system was kept switched off, since without a functioning flow control, it would have been useless anyway.

Most of the victims of the Gotthard road tunnel fire incident, where emergency exits to a parallel escape tunnel were available, died despite of having the possibility to escape, giving an example of irrational fatal human behavior. One driver even returned to his vehicle to pick up his mobile phone after he had already reached the safe shelter. That cost him his life.

In fact, the most sophisticated technical fire life safety measures are worthless, when humans don't react quickly and appropriately, by escaping immediately when being aware of an approaching smoke front.

---

[22] A list of worldwide serious tunnel fires can be found for instance in the PIARC publication [76].

Extensive media attention and subsequent political pressure after 1999 led to new national and international tunnel safety rules focusing on the following measures:

-   Providing twin-tube tunnels with unidirectional traffic
    when the traffic load exceeds a certain threshold level
-   Escape ways from tunnels to the outside in minimal distances
-   Fast, reliable incident and fire/smoke detection
-   Efficient traffic management systems in- and outside tunnels
-   Control of smoke propagation by control of longitudinal airflow and,
    in long tunnels with bidirectional traffic, concentrated extraction
-   Sewage systems for containment of spilled flammable liquids
-   other constructional measures, for instance lay-bys
-   Risk analysis as basis for conceptual decisions
-   Organizational and operational measures
-   Education of drivers, operators and emergency services

Those measures increased road tunnel safety significantly, and allowed for large funds to be invested, providing good business opportunities for the construction industry and equipment suppliers. Billions of Euros had been invested in refurbishments and new safety measures in many old and new European tunnels[23]. However, the lessons from the mentioned tunnel fires apply mostly to long single-tube tunnels with bidirectional traffic (category C). For highway tunnels with unidirectional traffic (category A), the situation is different.

The two described recent fires in 2018 occurred in modernized tunnels with state-of-the art ventilation systems with concentrated exhaust and flow control. At first sight, the systems worked as expected. Compared with the fires in 1999 and 2001, there was little damage, and all people could escape. But strictly speaking, both cases in 2018 were malfunctions, since the concentrated smoke extraction was operated at the wrong place, seriously affecting the smoke spread. Luckily, no fatalities had occurred, but it could have been worse.

---

[23] 380 Mio. EUR for the refurbishment of the Mont Blanc tunnel alone.

10 Closing remarks

More than 100 years ago, many long rail tunnels in the Alps had been equipped with powerful ventilation systems to dilute the smoke from the coal powered steam locomotives. After electrification of railways in the first half of the 20th century, those tunnel ventilation systems were not used anymore and had been removed.

Similarly, the situation in road tunnels has changed at the end of the 20th century, when vehicle emissions significantly dropped in most countries of the developed world due to strict air pollution control requirements. Before that, powerful transversal ventilation systems were required to limit exposure to toxic vehicle exhausts and achieve a sufficient visibility in long road tunnels. Today, the dilution of pollutants in and around road tunnels is rarely an issue[24]. For instance, the exhaust shaft in Fig. 20 had never been in operation. Fire ventilation has become the new design basis, even when fire disasters in road tunnels occur only rarely. Most road tunnel ventilation systems are hardly ever in operation.

There are many country-specific design standards and guidelines, where the requirements for the ventilation design and operation are more or less strictly defined; some are listed in the bibliography. However, a standard represents at most the 'state of the art' at the time when it was written, and therefore is obsolete when it becomes effective. Many standards and guidelines are not thought out in aspects of safety, cost-benefit efficiency, and practical application. Well-meant requirements may also result in unwanted side effects, which arise in practice. Few standards describe how to find and eliminate flaws and verify reliable accomplishment of goals.

In a legal investigation following an incident, the judges and lawyers consult whether the technical rules and standards have been followed. But who is responsible for the content of those standards? What about requirements that are detrimental to the safety of people and infrastructure? How far does the engineer's liability reach?

As a good example, in the NFPA standard [26] there is written: *'Nothing in this standard is intended to prevent the use of systems, methods, or devices of equivalent or superior quality, strength, fire resistance, effectiveness, durability, reliability, and safety over those prescribed by this standard, provided sufficient technical data demonstrates that the applied method material or device is equivalent to or superior to the requirements of this standard with respect to fire performance and safety.'*

Best practice should be based on logical reasoning and empirical evidence from practical experience.

---

[24] Exceptions are very long urban tunnels with sensitive environmental impact requirements, or tunnels in countries with a large part of vehicles without emission restrictions.

The following particular issues have been addressed in this Compendium:

- The usefulness of a ventilation concept shall be assessed by achievement of the goals described in chapter 2.3 and the guiding principle 'as simple as possible'. Tunnel ventilation is no 'rocket science'. The design provides only the basis for possible states of operation. What counts in practice is the reliable operational functionality of a ventilation system under traffic and different boundary conditions during its whole lifetime.
  For comparison: In road safety, your car's technical parameters on paper are less important than its actual condition and your driving abilities and appropriate fitness.

- Even appropriate, well thought-out ventilation systems may lead to failures with disastrous consequences, especially when not carefully implemented and thoroughly tested. That is often the case due to time and cost restrictions.
  Unlike your car, each tunnel is a unique construction.

- Critical velocity, which is a design value, must usually not be an operating state for fire ventilation in a tunnel. Most people died in tunnel fires due to fast smoke spread while the critical velocity was exceeded. In this respect, design and operation must be strictly distinguished. This important distinction has been neglected in some standards. The design case is not necessarily a required state of operation.
  Your car may be designed for a speed of 200 km/h. Does that mean that you always have to drive at that speed?

- Implementing a reliable flow control requires an elaborate technical effort for measuring devices, fan drives and controllers, which must be thoroughly tested and maintained. This results in increased complexity and costs. On the other side, the flow induced by the ventilation system in a tunnel without flow control is a random variable, depending on momentary boundary conditions. This may be reasonable for many applications, but not for fire ventilation, when human lives are at stake. Operating the ventilation system without precise, reliable flow measurement and control is like driving your car with closed eyes.

- Smoke exhaust for tunnels with unhindered unidirectional traffic is unnecessary, and may even increase the risk by possible faulty operation. By opening dampers ahead of the fire location, where cars are blocked, people may get trapped in the smoke. Without exhaust, but with a flow control limiting the air speed, they would have stayed safe. In fact, this has happened many times in real fire incidents and tests.

- High temperature resistance of jet fans is not only useless (as explained in chapter 8.9), but leads to additional technical problems. The requested temperature resistance of 400°C for jet fans in some European countries resulted in the development of new fan constructions and materials, which turned out to be unapt for road tunnel applications. For instance, the destroyed fan in Fig. 46 was optimized for high temperatures, but fell apart after a few years of operation. In the concerned tunnel, a fan blade crashed onto a bypassing car.

Humans err, technology does advance (and sometimes steps back), and circumstances may change. What is right now, may be obsolete tomorrow. We must not be reluctant to learn.
Look forward to the next edition.

I want to thank all my honorable clients and employers. They enabled me to do interesting work from which I derived my professional experience. Special thanks to many colleagues and peers, mechanical, electrical and civil engineers, safety experts, tunnel operators, fire fighters, systems and equipment providers and other professionals for useful hints and fruitful discussions. Namely to be mentioned are Dr. Volker Carstens and Dr. Bernd Hagenah who professionally reviewed the present book.

Finally, thanks to my family and friends for support and patience.

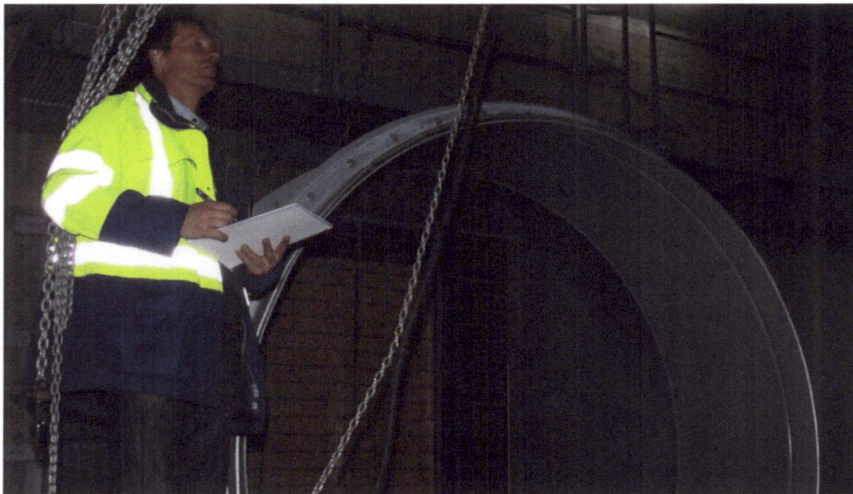

Fig. 73   The author supervising a fan installation

11 Bibliography

11.1 Norms, standards, guidelines

Since norms and guidelines are usually updated on a regular basis, no publication dates are indicated. It is advised to evaluate and consider the respective current version.

[1]   ASTRA 13001, Richtlinie Lüftung der Strassentunnel, CH

[2]   ASTRA 13002, Richtlinie Lüftung der Sicherheitsstollen von Strassentunneln, CH

[3]   ASTRA 13004, Branddetektion in Strassentunneln, CH

[4]   ASTRA 13011, Richtlinie Türen und Tore in Strassentunneln, CH

[5]   ASTRA 19004, Risikoanalyse für Tunnel der Nationalstrassen, CH

[6]   BD 78/99 Design Manual for Roads and Bridges, Design of Road Tunnels, GB

[7]   Directive 2004/54/EC on minimum safety requirements for tunnels in the Trans-European Road Network

[8]   Directive 2006/42/EC on machinery

[9]   Dossier Pilote Ventilation, CETU, F

[10] EN 12101-3 Smoke and heat control systems. Specification for powered smoke and heat control ventilators

[11] EN 12101-6 Smoke and heat control systems. Specification for pressure differential systems. Kits

[12] EN 50126 Railway applications – The specification and demonstration of Reliability, Availability, Maintainability and Safety (RAMS)

[13] IEC 61508 Functional safety (parts 1 - 7)

[14] ISO 12100, Safety of machinery: General principles for design, risk assessment and risk reduction

[15] ISO 12944 Paints & Varnishes - Corrosion Protection of Steel Structures by protective paint systems (parts 1 - 8)

[16] ISO 13347 Industrial fans - Determination of fan sound power levels under standardized laboratory conditions

[17] ISO 13350 Industrial fans - Performance testing of jet fans

[18] ISO 14694 Industrial fans – Specifications for balance quality and vibration levels

[19] ISO 1940 Mechanical vibration - Balance quality requirements for rotors in a constant (rigid) state

[20] ISO 39001 Road traffic safety (RTS) management systems — Requirements with guidance for use

[21] ISO 5801 Industrial fans - Performance testing using standardized airways

[22] ISO 5802 Industrial fans - Performance testing in situ

[23] Linee Guida per la progettazione della sicurezza nelle Gallerie Stradali, I

[24] Metodický pokyn, Větrání silničních tunelů, Volba systému, navrhování, provoz a zabezpečení jakosti větracích systémů silničních tunelů, CZ

[25] NFPA 130 Standard for Fixed Guideway Transit and Passenger Rail Systems, USA

[26] NFPA 502 Standard for Road Tunnels, Bridges, and Other Limited Access Highways, USA

[27] RABT, Richtlinien für die Ausstattung und den Betrieb von Straßentunneln, D

[28] RVS 09.02.31, Belüftung, A

[29] RVS 09.02.32, Luftbedarfsrechnung, A

[30] RVS 09.02.33, Lüftungsanlagen, Immissionsbelastung an Portalen, A

[31] RVS 09.03.11, Tunnel-Risikoanalysemodell, A

[32] SIA 197, Projektierung Tunnel, Grundlagen, CH

[33] SIA 197/2, Projektierung Tunnel, Strassentunnel, CH

[34] TP049, Technicke podmienky, Vetranie cestnych tunelov, SK

[35] TSI-SRT: REGULATION (EU) 2016/796, Technical specifications for interoperability relating to 'safety in railway tunnels'

[36] Vegtunneler Normaler Håndbok, N

11.2 Technical literature

[37] Ackeret J., Haerter A., Stahel M., Die Lüftung der Autotunnel, Bericht der Expertenkommission für Tunnellüftung an das Eidg. Amt für Strassen- und Flussbau, 1960

[38] Altenburger P., Riess I., Brandt R., Regelung der Luftströmung in Strassentunneln im Brandfall, ASTRA 2010/017, 2013

[39] BASt, Bewertung der Sicherheit von Straßentunneln, FE 03.0378/2004/FRB, 2007

[40] Beard A., Carvel R. (editors), The handbook of tunnel fire safety, 2nd ed. 2012

[41] Bernagaud C. et al., The treatment of air in road tunnels, state-of-the art of studies and works, CETU, 2010

[42] Betta V. et. al., Numerical Study of the Optimization of the Pitch Angle of an Alternative Jet Fan in a Longitudinal Tunnel Ventilation System, 2009

[43] Beyer M., Sturm P., Evaluation of Jet Fan Performance in Tunnels, 2016

[44] Bohl W., Elmendorf W., Strömungsmaschinen (1 und 2), 1980

[45] Bohl W., Technische Strömungslehre, Würzburg 1971

[46] Bopp R., Haag S., Peter A., Beschlagende Scheiben in Strassentunneln, 2004

[47] Brände in Verkehrstunneln - Bericht über Versuche im Maßstab 1:1; EUREKA-Projekt FIRETUN 499, 1998

[48] Design fires in road tunnels, NCHRP Synthesis 415, Transportation Research Board, 2011

[49] Dolejsky K., Pokorny W., Regelverhalten von Strassentunnellüftungen, 1984

[50] Drysdale D., An introduction to Fire Dynamics, 3rd ed. 2011

[51] Duffé P., Marec M., Cialdini P., Rapport Commun des missions administratives d'enquête technique Française et Italienne relative a la catastrophe survenue le 24 mars 1999 dans le tunnel du Mont Blanc, 1999

[52] Freibauer B. et al., Bemessungsgrundlagen für die Lüftung von Strassentunneln, 1978

[53] Haerter A. et al., Basic Principles for the Ventilation of Road Tunnels, prepared for the Swiss Federal Office of Highways, 1983

[54] Haerter A., Theoretische und experimentelle Untersuchungen über die Lüftungsanlagen von Strassentunneln, 1961

[55] Henschke H., Untersuchungen über verbesserte Strahlgebläse für Tunnellüftung, 1937

[56] Hu, L.H., Huo, R., Chow, W.K., Studies on buoyancy-driven back-layering flow in tunnel fires, 2008

[57] Idelchik I.E., Flow Resistance: A Design Guide for Engineers, 1989

[58] Ingason H., Li Y.Z., Technical Trade-offs using Fixed Fire Fighting Systems, 2014

[59] Ingason H., Lönnermark A., Li Y.Z., Model of ventilation flows during large tunnel fires, 2012

[60] Ingason H., Lönnermark A., Li Y.Z., Runehamar Tunnel Fire Tests, 2011

[61] Ingason H., Lönnermark A., Li Y.Z., Tunnel Fire Dynamics, 2015

[62] Jenssen G., Human Factor and Behavior: What we can learn from dealing with real road tunnel fires in Long Single-Bore Tunnels, 2015

[63] Kempf G., Einfluss der Wandeffekte auf die Treibstrahlwirkung eines Strahlgebläses, 1965

[64] Kubwimana, T., Salizzoni, P., Bergamini, E. et al., Wind-induced pressure at a tunnel portal, 2018

[65] Lönnermark A., On the Characteristics of Fires in Tunnels, 2005

[66] Massachussetts Highway Department, Memorial Tunnel Fire Ventilation Test Program, Comprehensive Test Report, 1995

[67] Mayer G., Brände in Straßentunneln: Abschätzung der Selbstrettungsmöglichkeiten der Tunnelnutzer mittels numerischer Rauchausbreitungssimulationen, 2006

[68] McPherson M., Subsurface Ventilation Engineering, 1993

[69] Meidinger U., Längslüftung von Autobahntunneln mit Strahlgebläsen, Schweizerische Bauzeitung, 1964

[70] Mellert L., Welte U. et. al., Elektromobilität und Tunnelsicherheit – Gefährdungen durch Elektrofahrzeugbrände, 2018

[71] Opstad K., Aune P., Henning J.E, Fire Emergency Ventilation Capacity for Road Tunnels with Considerable Slope, 1997

[72] Petrowski H., To Engineer is Human: The Role of Failure in Successful Design, 1985

[73] PIARC 05.04.B, 1995, Road Safety in Tunnels

[74] PIARC 05.05.B, 1999, Fire and smoke control in road tunnels

[75] PIARC 05.15.B, 2004, Traffic Incident Management Systems used in Road Tunnels

[76] PIARC 05.16.B, 2007, Systems and Equipment for fire and smoke control in road tunnels

[77] PIARC 2008R17, Human Factors and Road Tunnel Safety regarding Users

[78] PIARC 2010R1, Towards Development of a Risk Management Approach

[79] PIARC 2011R02, Road Tunnels: Operational Strategies for Emergency Ventilation

[80] PIARC 2012R12EN, Recommendations on management of maintenance and technical inspection of road tunnels

[81] PIARC 2012R14EN, Life cycle aspects of electrical road tunnel equipment,

[82] PIARC 2012R23EN, Current Practice for Risk Evaluation for Road Tunnels

[83] PIARC 2012R30EN, Social Acceptance of Risks and their Perception

[84] PIARC 2016R03EN, Fixed Fire Fighting Systems in Road Tunnels: Current Practices and Recommendations

[85] PIARC 2017R01EN, Design Fire Characteristics for Road Tunnels

[86] PIARC 2019R02EN, Road Tunnels: Vehicle Emissions and Air Demand for Ventilation

[87] PIARC 2019R03EN, Prevention and Mitigation of Tunnel Related Collisions

[88] PIARC 2019R05EN, Introduction to the RAMS Concept for Road Tunnel Operations

[89] Pokorny W. et al., Theoretische und praktische Untersuchungen zur Lüftung von Strassentunneln, 1978

[90] Pospisil P, Reducing Costs and Improving Safety of Road Tunnels, 2011

[91] Pospisil P., Ilg L. et al., Beeinflussung der Luftströmung in Strassentunneln im Brandfall, ASTRA 2007/002, 2010

[92] Pucher K., Die Selbstlüftung durch Fahrzeuge in einem Strassentunnel, 1979

[93] Purser, D.A, McAllister, J.L., Assessment of hazards to occupants from smoke, toxic gases and heat, SFPE Handbook of Fire Protection Engineering, 5th edition, 2016

[94] Riess I., Einsatz von Wassernebelanlagen bei Tunnelbränden - eine neue Bedingung für die Ereignislüftung, 2017

[95] Rohne E., Friction Losses of a Single Jet Due to its Contact with Vaulted Ceiling, 1991

[96] Rohne E., The Friction Losses on Walls Caused by a Row of Four Parallel Jet Flows, 1988

[97] Rohne E., Über die Längslüftung von Autotunneln mit Strahlventilatoren, 1964

[98] Rötzer H. et al., Einflüsse auf den Luftwechsel in Strassentunneln, 1986

[99] Singstad O., Ventilation of Vehicular Tunnels, 1929

[100] Sturm P., B. Höpperger B., Auswirkungen der Berücksichtigung der Temperaturerhöhung im Brandfall auf die Dimensionierung von quergelüfteten Straßentunneln, 2011

[101] Sturm P., B. Höpperger B., Betrachtung der Wärmefreisetzung im Brandfall, 2010

[102] Wagner W., Lufttechnische Anlagen, Würzburg 1997

[103] Wiesmann E., Die Ventilatoren: Berechnung, Entwurf und Anwendung, 1924

[104] Wiesmann E., Künstliche Lüftung im Stollen- und Tunnelbau sowie von Tunnels in Betrieb, 1919

[105] Winkler, M., Carvel, R., The effect of longitudinal ventilation on tenability during egress from passenger trains in tunnels during fire emergencies, 2015

[106] Winkler, M., Carvel, R., Ventilation and egress strategies for passenger train fires in tunnels, 2016

[107] Wirz W., Die Lüftung der Alpenstrassen-Tunnel, 1942

[108] Wouters P., Building Ventilation: The State of the Art, 2006

www.ingramcontent.com/pod-product-compliance
Lightning Source LLC
Chambersburg PA
CBHW041722210326
41598CB00007B/747

* 9 7 8 3 9 5 2 4 1 7 8 4 3 *